AF884581

Herd and Livestock Record Book

For Small Farm and Ranch Owners

Table Of Contents

This Record Book Belongs To

Location

Phone/ Contact

Email

Herd Profile

Animal Name / ID Tag ______________________________

Breed & Color / Markings __________________________

Date of Birth ____________________________________

Medication / Supplement ___________________________

____

Dosage ______________________________________

Frequency ____________________________________

Route (Oral / IM / SQ / Topical

Start Date: __________

End Date: __________

Notes ___________________

____

____

____

____

____

____

Animal Photo

Herd Profile

Animal Name / ID Tag ______________________________

Breed & Color / Markings __________________________

Date of Birth ____________________________________

Medication / Supplement ___________________________

____

Dosage ____

Frequency ____

Route (Oral / IM / SQ / Topical

Start Date: ___________

End Date: ___________

Notes ____________________

____

____

____

____

____

____

Animal Photo

Herd Profile

Animal Name / ID Tag ________________________________

Breed & Color / Markings ____________________________

Date of Birth _______________________________________

Medication / Supplement _____________________________

____

Dosage ____

Frequency ____

Route (Oral / IM / SQ / Topical

Start Date: ___________

End Date: ___________

Notes ____________________

____

____

____

____

____

____

Animal Photo

Herd Profile

Animal Name / ID Tag ______________________________

Breed & Color / Markings ______________________________

Date of Birth ______________________________

Medication / Supplement ______________________________

____

Dosage ______________________________

Frequency ______________________________

Route (Oral / IM / SQ / Topical

Start Date: ____________

End Date: ____________

Notes ____________________

____

____

____

____

____

____

Animal Photo

Herd Profile

Animal Name / ID Tag ______________________________

Breed & Color / Markings __________________________

Date of Birth ____________________________________

Medication / Supplement ___________________________

____

Dosage ____

Frequency _______________________________________

Route (Oral / IM / SQ / Topical

Start Date: ___________

End Date: ___________

Notes ____________________

____

____

____

____

____

____

Animal Photo

Herd Profile

Animal Name / ID Tag ______________________________

Breed & Color / Markings __________________________

Date of Birth _____________________________________

Medication / Supplement ___________________________

____

Dosage ____

Frequency ____

Route (Oral / IM / SQ / Topical

Start Date: ___________

End Date: ___________

Notes ____________________

____

____

____

____

____

____

Animal Photo

Herd Profile

Animal Name / ID Tag ______________________________

Breed & Color / Markings __________________________

Date of Birth ____________________________________

Medication / Supplement ___________________________

____

Dosage ____

Frequency _______________________________________

Route (Oral / IM / SQ / Topical

Start Date: __________

End Date: __________

Notes ___________________

____

____

____

____

____

____

Animal Photo

Herd Profile

Animal Name / ID Tag ________________________________

Breed & Color / Markings ____________________________

Date of Birth ______________________________________

Medication / Supplement ____________________________

____

Dosage ____

Frequency ____

Route (Oral / IM / SQ / Topical

Start Date: ___________

End Date: ___________

Notes ____________________

____

____

____

____

____

____

Animal Photo

Herd Profile

Animal Name / ID Tag ______________________________

Breed & Color / Markings __________________________

Date of Birth ____________________________________

Medication / Supplement ___________________________

____

Dosage ____

Frequency _______________________________________

Route (Oral / IM / SQ / Topical

Start Date: ___________

End Date: ___________

Notes ___________________

____

____

____

____

____

____

Animal Photo

Herd Profile

Animal Name / ID Tag ______________________________

Breed & Color / Markings __________________________

Date of Birth ____________________________________

Medication / Supplement ___________________________

____

Dosage ____

Frequency _______________________________________

Route (Oral / IM / SQ / Topical

Start Date: ___________

End Date: ___________

Notes ____________________

____

____

____

____

____

____

Animal Photo

Herd Profile

Animal Name / ID Tag ____________________________________

Breed & Color / Markings ________________________________

Date of Birth ____

Medication / Supplement __________________________________

____

Dosage ____

Frequency ____

Route (Oral / IM / SQ / Topical

Start Date: ____________

End Date: ____________

Notes ______________________

____

____

____

____

____

____

Animal Photo

Herd Profile

Animal Name / ID Tag ______________________________

Breed & Color / Markings ___________________________

Date of Birth _____________________________________

Medication / Supplement ____________________________

____

Dosage ____

Frequency ____

Route (Oral / IM / SQ / Topical

Start Date: ___________

End Date: ___________

Notes ____________________

____

____

____

____

____

____

Animal Photo

Herd Profile

Animal Name / ID Tag ______________________________

Breed & Color / Markings __________________________

Date of Birth ____________________________________

Medication / Supplement __________________________

____

Dosage ____

Frequency _______________________________________

Route (Oral / IM / SQ / Topical

Start Date: ____________

End Date: ____________

Notes ____________________

____

____

____

____

____

____

Animal Photo

Herd Profile

Animal Name / ID Tag ______________________________

Breed & Color / Markings __________________________

Date of Birth ____________________________________

Medication / Supplement ___________________________

____

Dosage ____

Frequency _______________________________________

Route (Oral / IM / SQ / Topical

Start Date: ___________

End Date: ___________

Notes ____________________

____

____

____

____

____

____

Animal Photo

Herd Profile

Animal Name / ID Tag ______________________________

Breed & Color / Markings __________________________

Date of Birth ____________________________________

Medication / Supplement ___________________________

____

Dosage ____

Frequency _____________________________________

Route (Oral / IM / SQ / Topical

Start Date: ___________

End Date: ___________

Notes ___________________

____

____

____

____

____

____

Animal Photo

Herd Profile

Animal Name / ID Tag ______________________________

Breed & Color / Markings __________________________

Date of Birth _____________________________________

Medication / Supplement ___________________________

____

Dosage ____

Frequency ____

Route (Oral / IM / SQ / Topical

Start Date: ___________

End Date: ___________

Notes ____________________

____

____

____

____

____

____

Animal Photo

Herd Profile

Animal Name / ID Tag ______________________________

Breed & Color / Markings __________________________

Date of Birth ____________________________________

Medication / Supplement ___________________________

____

Dosage ____

Frequency ____

Route (Oral / IM / SQ / Topical

Start Date: __________

End Date: __________

Notes ____________________

____

____

____

____

____

____

Animal Photo

Herd Profile

Animal Name / ID Tag ________________________________

Breed & Color / Markings ____________________________

Date of Birth ______________________________________

Medication / Supplement ____________________________

____

Dosage ____

Frequency ____

Route (Oral / IM / SQ / Topical

Start Date: ____________

End Date: ____________

Notes ____________________

____

____

____

____

____

____

Animal Photo

Herd Profile

Animal Name / ID Tag ______________________________

Breed & Color / Markings __________________________

Date of Birth ____________________________________

Medication / Supplement ___________________________

____

Dosage ____

Frequency ____

Route (Oral / IM / SQ / Topical

Start Date: ___________

End Date: ___________

Notes ___________________

____

____

____

____

____

____

Animal Photo

Herd Profile

Animal Name / ID Tag ______________________________

Breed & Color / Markings __________________________

Date of Birth ____________________________________

Medication / Supplement ___________________________

____

Dosage ____

Frequency ____

Route (Oral / IM / SQ / Topical

Start Date: ___________

End Date: ___________

Notes ___________________

____

____

____

____

____

____

Animal Photo

Herd Profile

Animal Name / ID Tag ______________________________

Breed & Color / Markings __________________________

Date of Birth ____________________________________

Medication / Supplement ___________________________

____

Dosage ____

Frequency ____

Route (Oral / IM / SQ / Topical

Start Date: ___________

End Date: ___________

Notes ___________________

____

____

____

____

____

____

Animal Photo

Herd Profile

Animal Name / ID Tag ______________________________

Breed & Color / Markings __________________________

Date of Birth ____________________________________

Medication / Supplement ___________________________

____

Dosage ____

Frequency _______________________________________

Route (Oral / IM / SQ / Topical

Start Date: ___________

End Date: ___________

Notes ____________________

____

____

____

____

____

____

Animal Photo

Herd Profile

Animal Name / ID Tag ______________________________

Breed & Color / Markings __________________________

Date of Birth ____________________________________

Medication / Supplement ___________________________

____

Dosage ____

Frequency _______________________________________

Route (Oral / IM / SQ / Topical

Start Date: ___________

End Date: ___________

Notes ___________________

____

____

____

____

____

____

Animal Photo

Herd Profile

Animal Name / ID Tag ______________________________

Breed & Color / Markings __________________________

Date of Birth ___________________________________

Medication / Supplement ___________________________

____

Dosage _______________________________________

Frequency _____________________________________

Route (Oral / IM / SQ / Topical

Start Date: __________

End Date: __________

Notes ___________________

____

____

____

____

____

____

Animal Photo

Herd Profile

Animal Name / ID Tag ______________________________

Breed & Color / Markings __________________________

Date of Birth _____________________________________

Medication / Supplement ___________________________

____

Dosage ____

Frequency ____

Route (Oral / IM / SQ / Topical

Start Date: __________

End Date: __________

Notes ___________________

____

____

____

____

____

____

Animal Photo

Herd Profile

Animal Name / ID Tag ________________________________

Breed & Color / Markings ____________________________

Date of Birth _______________________________________

Medication / Supplement ______________________________

____

Dosage ____

Frequency ____

Route (Oral / IM / SQ / Topical

Start Date: ___________

End Date: ___________

Notes _____________________

____

____

____

____

____

____

Animal Photo

Herd Profile

Animal Name / ID Tag ______________________________

Breed & Color / Markings __________________________

Date of Birth ____________________________________

Medication / Supplement ___________________________

____

Dosage ____

Frequency ____

Route (Oral / IM / SQ / Topical

Start Date: __________

End Date: __________

Notes ____________________

____

____

____

____

____

____

Animal Photo

Herd Profile

Animal Name / ID Tag ________________________________

Breed & Color / Markings ____________________________

Date of Birth ______________________________________

Medication / Supplement _____________________________

____

Dosage ____

Frequency ____

Route (Oral / IM / SQ / Topical

Start Date: ___________

End Date: ___________

Notes ____________________

____

____

____

____

____

____

Animal Photo

Herd Profile

Animal Name / ID Tag ________________________________

Breed & Color / Markings ____________________________

Date of Birth ______________________________________

Medication / Supplement _____________________________

____

Dosage ____

Frequency ____

Route (Oral / IM / SQ / Topical

Start Date: ___________

End Date: ___________

Notes ____________________

____

____

____

____

____

____

Animal Photo

Herd Profile

Animal Name / ID Tag ______________________________

Breed & Color / Markings __________________________

Date of Birth ____________________________________

Medication / Supplement ___________________________

____

Dosage ____

Frequency _______________________________________

Route (Oral / IM / SQ / Topical

Start Date: ___________

End Date: ___________

Notes ____________________

____

____

____

____

____

____

Animal Photo

Herd Profile

Animal Name / ID Tag ______________________________

Breed & Color / Markings __________________________

Date of Birth ____________________________________

Medication / Supplement ___________________________

____

Dosage ____

Frequency _______________________________________

Route (Oral / IM / SQ / Topical

Start Date: __________

End Date: __________

Notes ___________________

____

____

____

____

____

____

Animal Photo

Herd Profile

Animal Name / ID Tag ______________________________

Breed & Color / Markings __________________________

Date of Birth ____________________________________

Medication / Supplement ___________________________

____

Dosage ____

Frequency ____

Route (Oral / IM / SQ / Topical

Start Date: ___________

End Date: ___________

Notes ____________________

____

____

____

____

____

____

Animal Photo

Herd Profile

Animal Name / ID Tag ______________________________

Breed & Color / Markings __________________________

Date of Birth ____________________________________

Medication / Supplement ___________________________

____

Dosage ____

Frequency ______________________________________

Route (Oral / IM / SQ / Topical

Start Date: ___________

End Date: ___________

Notes ____________________

____

____

____

____

____

____

Animal Photo

Herd Profile

Animal Name / ID Tag ______________________________

Breed & Color / Markings __________________________

Date of Birth ____________________________________

Medication / Supplement ___________________________

____

Dosage ____

Frequency _______________________________________

Route (Oral / IM / SQ / Topical

Start Date: ___________

End Date: ___________

Notes ___________________

____

____

____

____

____

____

Animal Photo

Herd Profile

Animal Name / ID Tag ______________________________

Breed & Color / Markings __________________________

Date of Birth _____________________________________

Medication / Supplement ___________________________

____

Dosage ____

Frequency ____

Route (Oral / IM / SQ / Topical

Start Date: ___________

End Date: ___________

Notes ____________________

____

____

____

____

____

____

Animal Photo

Herd Profile

Animal Name / ID Tag ______________________________

Breed & Color / Markings __________________________

Date of Birth ____________________________________

Medication / Supplement ___________________________

____

Dosage ____

Frequency ______________________________________

Route (Oral / IM / SQ / Topical

Start Date: ___________

End Date: ___________

Notes ___________________

____

____

____

____

____

____

Animal Photo

Herd Profile

Animal Name / ID Tag ______________________________

Breed & Color / Markings __________________________

Date of Birth ____________________________________

Medication / Supplement ___________________________

____

Dosage ____

Frequency ____

Route (Oral / IM / SQ / Topical

Start Date: ___________

End Date: ___________

Notes ___________________

____

____

____

____

____

____

Animal Photo

Herd Profile

Animal Name / ID Tag ________________________________

Breed & Color / Markings ____________________________

Date of Birth ______________________________________

Medication / Supplement ____________________________

____

Dosage ____

Frequency ____

Route (Oral / IM / SQ / Topical

Start Date: ___________

End Date: ___________

Notes ____________________

____

____

____

____

____

____

Animal Photo

Herd Profile

Animal Name / ID Tag ______________________________

Breed & Color / Markings __________________________

Date of Birth ____________________________________

Medication / Supplement ___________________________

____

Dosage ____

Frequency ______________________________________

Route (Oral / IM / SQ / Topical

Start Date: ___________

End Date: ___________

Notes ___________________

____

____

____

____

____

____

Animal Photo

Herd Profile

Animal Name / ID Tag ______________________________

Breed & Color / Markings __________________________

Date of Birth ____________________________________

Medication / Supplement ___________________________

____

Dosage ____

Frequency ____

Route (Oral / IM / SQ / Topical

Start Date: ___________

End Date: ___________

Notes ____________________

____

____

____

____

____

____

Animal Photo

Herd Profile

Animal Name / ID Tag ______________________________

Breed & Color / Markings __________________________

Date of Birth _____________________________________

Medication / Supplement ___________________________

____

Dosage ____

Frequency _______________________________________

Route (Oral / IM / SQ / Topical

Start Date: ___________

End Date: ___________

Notes ___________________

____

____

____

____

____

____

Animal Photo

Herd Profile

Animal Name / ID Tag ______________________________

Breed & Color / Markings __________________________

Date of Birth ____________________________________

Medication / Supplement ___________________________

____

Dosage ____

Frequency _______________________________________

Route (Oral / IM / SQ / Topical

Start Date: ___________

End Date: ___________

Notes ____________________

____

____

____

____

____

____

Animal Photo

Herd Profile

Animal Name / ID Tag ________________________________

Breed & Color / Markings ____________________________

Date of Birth _______________________________________

Medication / Supplement _____________________________

____

Dosage ____

Frequency ____

Route (Oral / IM / SQ / Topical

Start Date: __________

End Date: __________

Notes ____________________

____

____

____

____

____

____

Animal Photo

Herd Profile

Animal Name / ID Tag ______________________________

Breed & Color / Markings __________________________

Date of Birth ____________________________________

Medication / Supplement ___________________________

____

Dosage ____

Frequency ______________________________________

Route (Oral / IM / SQ / Topical

Start Date: ___________

End Date: ___________

Notes ____________________

____

____

____

____

____

____

Animal Photo

Herd Profile

Animal Name / ID Tag ________________________________

Breed & Color / Markings ____________________________

Date of Birth ______________________________________

Medication / Supplement _____________________________

____

Dosage ____

Frequency ____

Route (Oral / IM / SQ / Topical

Start Date: ____________

End Date: ____________

Notes ____________________

____

____

____

____

____

____

Animal Photo

Herd Profile

Animal Name / ID Tag ______________________________

Breed & Color / Markings __________________________

Date of Birth ____________________________________

Medication / Supplement ___________________________

____

Dosage ____

Frequency ______________________________________

Route (Oral / IM / SQ / Topical

Start Date: ___________

End Date: ___________

Notes ___________________

____

____

____

____

____

____

Animal Photo

Herd Profile

Animal Name / ID Tag ______________________________

Breed & Color / Markings __________________________

Date of Birth ____________________________________

Medication / Supplement __________________________

____

Dosage ____

Frequency ______________________________________

Route (Oral / IM / SQ / Topical

Start Date: __________

End Date: __________

Notes ___________________

____

____

____

____

____

____

Animal Photo

Herd Profile

Animal Name / ID Tag ______________________________

Breed & Color / Markings __________________________

Date of Birth ____________________________________

Medication / Supplement ___________________________

____

Dosage ____

Frequency ______________________________________

Route (Oral / IM / SQ / Topical

Start Date: __________

End Date: __________

Notes ____________________

____

____

____

____

____

____

Animal Photo

Herd Profile

Animal Name / ID Tag ______________________________

Breed & Color / Markings __________________________

Date of Birth ____________________________________

Medication / Supplement ___________________________

____

Dosage ____

Frequency ____

Route (Oral / IM / SQ / Topical

Start Date: __________

End Date: __________

Notes ___________________

____

____

____

____

____

____

Animal Photo

Herd Profile

Animal Name / ID Tag ______________________________

Breed & Color / Markings __________________________

Date of Birth ____________________________________

Medication / Supplement ___________________________

____

Dosage ____

Frequency _______________________________________

Route (Oral / IM / SQ / Topical

Start Date: ___________

End Date: ___________

Notes ___________________

____

____

____

____

____

____

Animal Photo

Daily Medication Log

Date	Animal Name/ ID	Medication/ Supplement	Dose	Route	Time/ Initials

Daily Medication Log

Date	Animal Name/ ID	Medication/ Supplement	Dose	Route	Time/ Initials

Daily Medication Log

Date	Animal Name/ ID	Medication/ Supplement	Dose	Route	Time/ Initials

Daily Medication Log

Date	Animal Name/ ID	Medication/ Supplement	Dose	Route	Time/ Initials

Daily Medication Log

Date	Animal Name/ ID	Medication/ Supplement	Dose	Route	Time/ Initials

Daily Medication Log

Date	Animal Name/ ID	Medication/ Supplement	Dose	Route	Time/ Initials

Daily Medication Log

Date	Animal Name/ ID	Medication/ Supplement	Dose	Route	Time/ Initials

Daily Medication Log

Date	Animal Name/ ID	Medication/ Supplement	Dose	Route	Time/ Initials

Vaccination Record

Animal Name / ID ________________________________

Breed ________________________________

Date of Birth ________________________________

Vaccine	Lot #	Batch #	Date Given	Adminstered By

Booster Due	Lot #	Batch #	Date Given	Adminstered By

Dewormed	Lot #	Batch #	Date Given	Adminstered By

Notes ________________________________

____

____

____

Vaccination Record

Animal Name / ID ________________________________

Breed ________________________________

Date of Birth ________________________________

Vaccine	Lot #	Batch #	Date Given	Adminstered By

Booster Due	Lot #	Batch #	Date Given	Adminstered By

Dewormed	Lot #	Batch #	Date Given	Adminstered By

Notes ________________________________

____

____

____

Vaccination Record

Animal Name / ID ______________________________

Breed ______________________________

Date of Birth ______________________________

Vaccine	Lot #	Batch #	Date Given	Adminstered By

Booster Due	Lot #	Batch #	Date Given	Adminstered By

Dewormed	Lot #	Batch #	Date Given	Adminstered By

Notes ______________________________

____

____

____

Vaccination Record

Animal Name / ID ______________________________

Breed ______________________________

Date of Birth ______________________________

Vaccine	Lot #	Batch #	Date Given	Adminstered By

Booster Due	Lot #	Batch #	Date Given	Adminstered By

Dewormed	Lot #	Batch #	Date Given	Adminstered By

Notes ______________________________

____

____

____

Vaccination Record

Animal Name / ID ______________________________

Breed ______________________________

Date of Birth ______________________________

Vaccine	Lot #	Batch #	Date Given	Adminstered By

Booster Due	Lot #	Batch #	Date Given	Adminstered By

Dewormed	Lot #	Batch #	Date Given	Adminstered By

Notes ______________________________

____

____

____

Vaccination Record

Animal Name / ID ______________________________

Breed ______________________________

Date of Birth ______________________________

Vaccine	Lot #	Batch #	Date Given	Adminstered By

Booster Due	Lot #	Batch #	Date Given	Adminstered By

Dewormed	Lot #	Batch #	Date Given	Adminstered By

Notes ______________________________

____

____

____

Vaccination Record

Animal Name / ID ______________________________

Breed ______________________________

Date of Birth ______________________________

Vaccine	Lot #	Batch #	Date Given	Adminstered By

Booster Due	Lot #	Batch #	Date Given	Adminstered By

Dewormed	Lot #	Batch #	Date Given	Adminstered By

Notes ______________________________

____

____

____

Vaccination Record

Animal Name / ID ________________________________

Breed ________________________________

Date of Birth ________________________________

Vaccine	Lot #	Batch #	Date Given	Adminstered By

Booster Due	Lot #	Batch #	Date Given	Adminstered By

Dewormed	Lot #	Batch #	Date Given	Adminstered By

Notes ________________________________

____

____

____

Vaccination Record

Animal Name / ID ______________________________

Breed ______________________________

Date of Birth ______________________________

Vaccine	Lot #	Batch #	Date Given	Adminstered By

Booster Due	Lot #	Batch #	Date Given	Adminstered By

Dewormed	Lot #	Batch #	Date Given	Adminstered By

Notes ______________________________

____

____

____

Vaccination Record

Animal Name / ID ______________________________

Breed ______________________________

Date of Birth ______________________________

Vaccine	Lot #	Batch #	Date Given	Adminstered By

Booster Due	Lot #	Batch #	Date Given	Adminstered By

Dewormed	Lot #	Batch #	Date Given	Adminstered By

Notes ______________________________

____

____

____

Vaccination Record

Animal Name / ID ______________________________

Breed ______________________________

Date of Birth ______________________________

Vaccine	Lot #	Batch #	Date Given	Adminstered By

Booster Due	Lot #	Batch #	Date Given	Adminstered By

Dewormed	Lot #	Batch #	Date Given	Adminstered By

Notes ______________________________

____

____

____

Vaccination Record

Animal Name / ID ________________________________

Breed ________________________________

Date of Birth ________________________________

Vaccine	Lot #	Batch #	Date Given	Adminstered By

Booster Due	Lot #	Batch #	Date Given	Adminstered By

Dewormed	Lot #	Batch #	Date Given	Adminstered By

Notes ________________________________

____

____

____

Vaccination Record

Animal Name / ID ________________________________

Breed ________________________________

Date of Birth ________________________________

Vaccine	Lot #	Batch #	Date Given	Adminstered By

Booster Due	Lot #	Batch #	Date Given	Adminstered By

Dewormed	Lot #	Batch #	Date Given	Adminstered By

Notes ________________________________

____

____

____

Vaccination Record

Animal Name / ID ______________________________

Breed ______________________________

Date of Birth ______________________________

Vaccine	Lot #	Batch #	Date Given	Adminstered By

Booster Due	Lot #	Batch #	Date Given	Adminstered By

Dewormed	Lot #	Batch #	Date Given	Adminstered By

Notes ______________________________

____

____

____

Vaccination Record

Animal Name / ID ______________________________

Breed ______________________________

Date of Birth ______________________________

Vaccine	Lot #	Batch #	Date Given	Adminstered By

Booster Due	Lot #	Batch #	Date Given	Adminstered By

Dewormed	Lot #	Batch #	Date Given	Adminstered By

Notes______________________________

____

____

____

Vaccination Record

Animal Name / ID ______________________________

Breed ______________________________________

Date of Birth _________________________________

Vaccine	Lot #	Batch #	Date Given	Adminstered By

Booster Due	Lot #	Batch #	Date Given	Adminstered By

Dewormed	Lot #	Batch #	Date Given	Adminstered By

Notes____

____

____

____

Vaccination Record

Animal Name / ID ______________________________

Breed ______________________________

Date of Birth ______________________________

Vaccine	Lot #	Batch #	Date Given	Adminstered By

Booster Due	Lot #	Batch #	Date Given	Adminstered By

Dewormed	Lot #	Batch #	Date Given	Adminstered By

Notes ______________________________

____

____

____

Vaccination Record

Animal Name / ID ________________________________

Breed ____

Date of Birth __________________________________

Vaccine	Lot #	Batch #	Date Given	Adminstered By

Booster Due	Lot #	Batch #	Date Given	Adminstered By

Dewormed	Lot #	Batch #	Date Given	Adminstered By

Notes ____

____

____

____

Vaccination Record

Animal Name / ID ________________________________

Breed ________________________________

Date of Birth ________________________________

Vaccine	Lot #	Batch #	Date Given	Adminstered By

Booster Due	Lot #	Batch #	Date Given	Adminstered By

Dewormed	Lot #	Batch #	Date Given	Adminstered By

Notes ________________________________

____

____

____

Vaccination Record

Animal Name / ID ______________________________

Breed ______________________________

Date of Birth ______________________________

Vaccine	Lot #	Batch #	Date Given	Adminstered By

Booster Due	Lot #	Batch #	Date Given	Adminstered By

Dewormed	Lot #	Batch #	Date Given	Adminstered By

Notes ______________________________

____

____

____

Vaccination Record

Animal Name / ID ______________________________

Breed ______________________________

Date of Birth ______________________________

Vaccine	Lot #	Batch #	Date Given	Adminstered By

Booster Due	Lot #	Batch #	Date Given	Adminstered By

Dewormed	Lot #	Batch #	Date Given	Adminstered By

Notes ______________________________

____

____

____

Vaccination Record

Animal Name / ID ________________________________

Breed ________________________________

Date of Birth ________________________________

Vaccine	Lot #	Batch #	Date Given	Adminstered By

Booster Due	Lot #	Batch #	Date Given	Adminstered By

Dewormed	Lot #	Batch #	Date Given	Adminstered By

Notes ________________________________

____

____

____

Vaccination Record

Animal Name / ID ______________________________

Breed ______________________________

Date of Birth ______________________________

Vaccine	Lot #	Batch #	Date Given	Adminstered By

Booster Due	Lot #	Batch #	Date Given	Adminstered By

Dewormed	Lot #	Batch #	Date Given	Adminstered By

Notes ______________________________

____

____

____

Vaccination Record

Animal Name / ID ________________________________

Breed ________________________________

Date of Birth ________________________________

Vaccine	Lot #	Batch #	Date Given	Adminstered By

Booster Due	Lot #	Batch #	Date Given	Adminstered By

Dewormed	Lot #	Batch #	Date Given	Adminstered By

Notes ________________________________

____

____

____

Vaccination Record

Animal Name / ID ______________________________

Breed ______________________________

Date of Birth ______________________________

Vaccine	Lot #	Batch #	Date Given	Adminstered By

Booster Due	Lot #	Batch #	Date Given	Adminstered By

Dewormed	Lot #	Batch #	Date Given	Adminstered By

Notes ______________________________

____

____

____

Vaccination Record

Animal Name / ID ________________________________

Breed ________________________________

Date of Birth ________________________________

Vaccine	Lot #	Batch #	Date Given	Adminstered By

Booster Due	Lot #	Batch #	Date Given	Adminstered By

Dewormed	Lot #	Batch #	Date Given	Adminstered By

Notes ________________________________

____

____

____

Vaccination Record

Animal Name / ID ______________________________

Breed ______________________________

Date of Birth ______________________________

Vaccine	Lot #	Batch #	Date Given	Adminstered By

Booster Due	Lot #	Batch #	Date Given	Adminstered By

Dewormed	Lot #	Batch #	Date Given	Adminstered By

Notes ______________________________

____

____

____

Vaccination Record

Animal Name / ID ______________________________

Breed ______________________________________

Date of Birth ________________________________

Vaccine	Lot #	Batch #	Date Given	Adminstered By

Booster Due	Lot #	Batch #	Date Given	Adminstered By

Dewormed	Lot #	Batch #	Date Given	Adminstered By

Notes ____

____

____

____

Vaccination Record

Animal Name / ID ______________________________

Breed ______________________________

Date of Birth ______________________________

Vaccine	Lot #	Batch #	Date Given	Adminstered By

Booster Due	Lot #	Batch #	Date Given	Adminstered By

Dewormed	Lot #	Batch #	Date Given	Adminstered By

Notes ______________________________

____

____

____

Vaccination Record

Animal Name / ID ____________________

Breed ____________________

Date of Birth ____________________

Vaccine	Lot #	Batch #	Date Given	Adminstered By

Booster Due	Lot #	Batch #	Date Given	Adminstered By

Dewormed	Lot #	Batch #	Date Given	Adminstered By

Notes ____________________

Vaccination Record

Animal Name / ID ________________________________

Breed ________________________________

Date of Birth ________________________________

Vaccine	Lot #	Batch #	Date Given	Adminstered By

Booster Due	Lot #	Batch #	Date Given	Adminstered By

Dewormed	Lot #	Batch #	Date Given	Adminstered By

Notes ________________________________

____

____

____

Vaccination Record

Animal Name / ID ______________________________

Breed ______________________________________

Date of Birth ________________________________

Vaccine	Lot #	Batch #	Date Given	Adminstered By

Booster Due	Lot #	Batch #	Date Given	Adminstered By

Dewormed	Lot #	Batch #	Date Given	Adminstered By

Notes____

____

____

____

Vaccination Record

Animal Name / ID ______________________________

Breed ______________________________

Date of Birth ______________________________

Vaccine	Lot #	Batch #	Date Given	Adminstered By

Booster Due	Lot #	Batch #	Date Given	Adminstered By

Dewormed	Lot #	Batch #	Date Given	Adminstered By

Notes ______________________________

____

____

____

Vaccination Record

Animal Name / ID ______________________________

Breed ______________________________

Date of Birth ______________________________

Vaccine	Lot #	Batch #	Date Given	Adminstered By

Booster Due	Lot #	Batch #	Date Given	Adminstered By

Dewormed	Lot #	Batch #	Date Given	Adminstered By

Notes ______________________________

____

____

____

Vaccination Record

Animal Name / ID ______________________________

Breed ______________________________

Date of Birth ______________________________

Vaccine	Lot #	Batch #	Date Given	Adminstered By

Booster Due	Lot #	Batch #	Date Given	Adminstered By

Dewormed	Lot #	Batch #	Date Given	Adminstered By

Notes ______________________________

____

____

____

Vaccination Record

Animal Name / ID ______________________________

Breed ______________________________

Date of Birth ______________________________

Vaccine	Lot #	Batch #	Date Given	Adminstered By

Booster Due	Lot #	Batch #	Date Given	Adminstered By

Dewormed	Lot #	Batch #	Date Given	Adminstered By

Notes ______________________________

____

____

____

Vaccination Record

Animal Name / ID ______________________________

Breed ______________________________

Date of Birth ______________________________

Vaccine	Lot #	Batch #	Date Given	Adminstered By

Booster Due	Lot #	Batch #	Date Given	Adminstered By

Dewormed	Lot #	Batch #	Date Given	Adminstered By

Notes ______________________________

____

____

____

Vaccination Record

Animal Name / ID ________________________________

Breed ________________________________

Date of Birth ________________________________

Vaccine	Lot #	Batch #	Date Given	Adminstered By

Booster Due	Lot #	Batch #	Date Given	Adminstered By

Dewormed	Lot #	Batch #	Date Given	Adminstered By

Notes ________________________________

____

____

____

Vaccination Record

Animal Name / ID ________________________________

Breed ________________________________

Date of Birth ________________________________

Vaccine	Lot #	Batch #	Date Given	Adminstered By

Booster Due	Lot #	Batch #	Date Given	Adminstered By

Dewormed	Lot #	Batch #	Date Given	Adminstered By

Notes ________________________________

____

____

____

Vaccination Record

Animal Name / ID ______________________________

Breed ______________________________________

Date of Birth _______________________________

Vaccine	Lot #	Batch #	Date Given	Adminstered By

Booster Due	Lot #	Batch #	Date Given	Adminstered By

Dewormed	Lot #	Batch #	Date Given	Adminstered By

Notes ____

____

____

____

Vaccination Record

Animal Name / ID ______________________________

Breed ____

Date of Birth _________________________________

Vaccine	Lot #	Batch #	Date Given	Adminstered By

Booster Due	Lot #	Batch #	Date Given	Adminstered By

Dewormed	Lot #	Batch #	Date Given	Adminstered By

Notes ____

____

____

____

Vaccination Record

Animal Name / ID ______________________________

Breed ______________________________

Date of Birth ______________________________

Vaccine	Lot #	Batch #	Date Given	Adminstered By

Booster Due	Lot #	Batch #	Date Given	Adminstered By

Dewormed	Lot #	Batch #	Date Given	Adminstered By

Notes ______________________________

____

____

____

Vaccination Record

Animal Name / ID ______________________________

Breed ______________________________

Date of Birth ______________________________

Vaccine	Lot #	Batch #	Date Given	Adminstered By

Booster Due	Lot #	Batch #	Date Given	Adminstered By

Dewormed	Lot #	Batch #	Date Given	Adminstered By

Notes ______________________________

____

____

____

Vaccination Record

Animal Name / ID ________________________________

Breed ________________________________

Date of Birth ________________________________

Vaccine	Lot #	Batch #	Date Given	Adminstered By

Booster Due	Lot #	Batch #	Date Given	Adminstered By

Dewormed	Lot #	Batch #	Date Given	Adminstered By

Notes ________________________________

____

____

____

Vaccination Record

Animal Name / ID ______________________________

Breed ______________________________

Date of Birth ______________________________

Vaccine	Lot #	Batch #	Date Given	Adminstered By

Booster Due	Lot #	Batch #	Date Given	Adminstered By

Dewormed	Lot #	Batch #	Date Given	Adminstered By

Notes ______________________________

____

____

____

Vaccination Record

Animal Name / ID ______________________________

Breed ______________________________

Date of Birth ______________________________

Vaccine	Lot #	Batch #	Date Given	Adminstered By

Booster Due	Lot #	Batch #	Date Given	Adminstered By

Dewormed	Lot #	Batch #	Date Given	Adminstered By

Notes ______________________________

____

____

____

Vaccination Record

Animal Name / ID ____________________________________

Breed ____

Date of Birth ____

Vaccine	Lot #	Batch #	Date Given	Adminstered By

Booster Due	Lot #	Batch #	Date Given	Adminstered By

Dewormed	Lot #	Batch #	Date Given	Adminstered By

Notes ____

____

____

____

Vaccination Record

Animal Name / ID ______________________________

Breed ____

Date of Birth _________________________________

Vaccine	Lot #	Batch #	Date Given	Adminstered By

Booster Due	Lot #	Batch #	Date Given	Adminstered By

Dewormed	Lot #	Batch #	Date Given	Adminstered By

Notes ____

____

____

____

Vaccination Record

Animal Name / ID ______________________________

Breed ______________________________

Date of Birth ______________________________

Vaccine	Lot #	Batch #	Date Given	Adminstered By

Booster Due	Lot #	Batch #	Date Given	Adminstered By

Dewormed	Lot #	Batch #	Date Given	Adminstered By

Notes ______________________________

____

____

____

Vaccination Record

Animal Name / ID ________________________________

Breed ________________________________

Date of Birth ________________________________

Vaccine	Lot #	Batch #	Date Given	Adminstered By

Booster Due	Lot #	Batch #	Date Given	Adminstered By

Dewormed	Lot #	Batch #	Date Given	Adminstered By

Notes ________________________________

____

____

____

Farrier / Hoof Care Record

Animal Name / ID ______________________________
Date ____________________
Farrier ____________________

Service Type (circle)
Trim / Full Set / Front Shoes / Reset / Corrective

Hoof Worked On (circle)
LF /RF / LH / RH / ALL

Next Visit Due ___________

Hoof Condition / Notes ______________________________
____
____
____
____
____
____

Farrier / Hoof Care Record

Animal Name / ID ________________________________

Date ____________________

Farrier ____________________

Service Type (circle)
Trim / Full Set / Front Shoes / Reset / Corrective

Hoof Worked On (circle)
LF /RF / LH / RH / ALL

Next Visit Due ___________

Hoof Condition / Notes ______________________________

____

____

____

____

____

____

Farrier / Hoof Care Record

Animal Name / ID ________________________________
Date ____________________
Farrier __________________

Service Type (circle)
Trim / Full Set / Front Shoes / Reset / Corrective

Hoof Worked On (circle)
LF /RF / LH / RH / ALL

Next Visit Due ___________

Hoof Condition / Notes ______________________________
____
____
____
____
____
____

Farrier / Hoof Care Record

Animal Name / ID ________________________________
Date ____________________
Farrier ___________________

Service Type (circle)
Trim / Full Set / Front Shoes / Reset / Corrective

Hoof Worked On (circle)
LF /RF / LH / RH / ALL

Next Visit Due ___________

Hoof Condition / Notes ______________________________
____
____
____
____
____
____

Farrier / Hoof Care Record

Animal Name / ID ________________________________

Date ____________________

Farrier __________________

Service Type (circle)
Trim / Full Set / Front Shoes / Reset / Corrective

Hoof Worked On (circle)
LF /RF / LH / RH / ALL

Next Visit Due ___________

Hoof Condition / Notes ______________________________

____

____

____

____

____

____

Farrier / Hoof Care Record

Animal Name / ID ________________________________

Date ____________________

Farrier __________________

Service Type (circle)
Trim / Full Set / Front Shoes / Reset / Corrective

Hoof Worked On (circle)
LF /RF / LH / RH / ALL

Next Visit Due ___________

Hoof Condition / Notes ______________________________

____

____

____

____

____

____

Farrier / Hoof Care Record

Animal Name / ID ______________________________

Date ____________________

Farrier __________________

Service Type (circle)
Trim / Full Set / Front Shoes / Reset / Corrective

Hoof Worked On (circle)
LF /RF / LH / RH / ALL

Next Visit Due ___________

Hoof Condition / Notes ______________________________

____

____

____

____

____

____

Farrier / Hoof Care Record

Animal Name / ID ________________________________
Date ____________________
Farrier ___________________

Service Type (circle)
Trim / Full Set / Front Shoes / Reset / Corrective

Hoof Worked On (circle)
LF /RF / LH / RH / ALL

Next Visit Due ___________

Hoof Condition / Notes ______________________________
____
____
____
____
____
____

Farrier / Hoof Care Record

Animal Name / ID ______________________________

Date ____________________

Farrier __________________

Service Type (circle)
Trim / Full Set / Front Shoes / Reset / Corrective

Hoof Worked On (circle)
LF /RF / LH / RH / ALL

Next Visit Due ___________

Hoof Condition / Notes ______________________________

____

____

____

____

____

____

Farrier / Hoof Care Record

Animal Name / ID ________________________________
Date ____________________
Farrier __________________

Service Type (circle)
Trim / Full Set / Front Shoes / Reset / Corrective

Hoof Worked On (circle)
LF /RF / LH / RH / ALL

Next Visit Due ___________

Hoof Condition / Notes ______________________________
____
____
____
____
____
____

Farrier / Hoof Care Record

Animal Name / ID ________________________________

Date ____________________

Farrier __________________

Service Type (circle)
Trim / Full Set / Front Shoes / Reset / Corrective

Hoof Worked On (circle)
LF /RF / LH / RH / ALL

Next Visit Due ___________

Hoof Condition / Notes ______________________________

____

____

____

____

____

____

Farrier / Hoof Care Record

Animal Name / ID ______________________________

Date ___________________

Farrier _________________

Service Type (circle)

Trim / Full Set / Front Shoes / Reset / Corrective

Hoof Worked On (circle)

LF /RF / LH / RH / ALL

Next Visit Due __________

Hoof Condition / Notes ____________________________

____

____

____

____

____

____

Farrier / Hoof Care Record

Animal Name / ID ________________________________
Date ____________________
Farrier __________________

Service Type (circle)
Trim / Full Set / Front Shoes / Reset / Corrective

Hoof Worked On (circle)
LF /RF / LH / RH / ALL

Next Visit Due ___________

Hoof Condition / Notes ______________________________
____
____
____
____
____
____

Farrier / Hoof Care Record

Animal Name / ID ________________________________

Date ____________________

Farrier __________________

Service Type (circle)
Trim / Full Set / Front Shoes / Reset / Corrective

Hoof Worked On (circle)
LF /RF / LH / RH / ALL

Next Visit Due ___________

Hoof Condition / Notes ______________________________

____

____

____

____

____

____

Farrier / Hoof Care Record

Animal Name / ID ______________________________
Date ___________________
Farrier _________________

Service Type (circle)
Trim / Full Set / Front Shoes / Reset / Corrective

Hoof Worked On (circle)
LF /RF / LH / RH / ALL

Next Visit Due __________

Hoof Condition / Notes ____________________________
____
____
____
____
____
____

Farrier / Hoof Care Record

Animal Name / ID ________________________________

Date ____________________

Farrier ___________________

Service Type (circle)
Trim / Full Set / Front Shoes / Reset / Corrective

Hoof Worked On (circle)
LF /RF / LH / RH / ALL

Next Visit Due ___________

Hoof Condition / Notes ______________________________

____

____

____

____

____

____

Farrier / Hoof Care Record

Animal Name / ID ________________________________

Date ____________________

Farrier __________________

Service Type (circle)
Trim / Full Set / Front Shoes / Reset / Corrective

Hoof Worked On (circle)
LF /RF / LH / RH / ALL

Next Visit Due ___________

Hoof Condition / Notes ______________________________

____

____

____

____

____

____

Farrier / Hoof Care Record

Animal Name / ID ______________________________

Date ____________________

Farrier __________________

Service Type (circle)
Trim / Full Set / Front Shoes / Reset / Corrective

Hoof Worked On (circle)
LF /RF / LH / RH / ALL

Next Visit Due ___________

Hoof Condition / Notes ______________________________

____

____

____

____

____

____

Farrier / Hoof Care Record

Animal Name / ID ______________________________

Date ____________________

Farrier ___________________

Service Type (circle)
Trim / Full Set / Front Shoes / Reset / Corrective

Hoof Worked On (circle)
LF /RF / LH / RH / ALL

Next Visit Due ___________

Hoof Condition / Notes ______________________________

____

____

____

____

____

____

Farrier / Hoof Care Record

Animal Name / ID ______________________________

Date ___________________

Farrier _________________

Service Type (circle)
Trim / Full Set / Front Shoes / Reset / Corrective

Hoof Worked On (circle)
LF /RF / LH / RH / ALL

Next Visit Due __________

Hoof Condition / Notes ____________________________

____

____

____

____

____

____

Farrier / Hoof Care Record

Animal Name / ID ________________________________

Date ____________________

Farrier ___________________

Service Type (circle)
Trim / Full Set / Front Shoes / Reset / Corrective

Hoof Worked On (circle)
LF /RF / LH / RH / ALL

Next Visit Due ___________

Hoof Condition / Notes ______________________________

____

____

____

____

____

____

Farrier / Hoof Care Record

Animal Name / ID ________________________________
Date ____________________
Farrier ___________________

Service Type (circle)
Trim / Full Set / Front Shoes / Reset / Corrective

Hoof Worked On (circle)
LF /RF / LH / RH / ALL

Next Visit Due ___________

Hoof Condition / Notes

____

____

____

____

____

____

____

Farrier / Hoof Care Record

Animal Name / ID ________________________________

Date ____________________

Farrier __________________

Service Type (circle)
Trim / Full Set / Front Shoes / Reset / Corrective

Hoof Worked On (circle)
LF /RF / LH / RH / ALL

Next Visit Due ___________

Hoof Condition / Notes ______________________________

____

____

____

____

____

____

Farrier / Hoof Care Record

Animal Name / ID ______________________________

Date ___________________

Farrier _________________

Service Type (circle)

Trim / Full Set / Front Shoes / Reset / Corrective

Hoof Worked On (circle)

LF /RF / LH / RH / ALL

Next Visit Due __________

Hoof Condition / Notes ____________________________

____

____

____

____

____

____

Farrier / Hoof Care Record

Animal Name / ID ________________________________

Date ____________________

Farrier ____________________

Service Type (circle)
Trim / Full Set / Front Shoes / Reset / Corrective

Hoof Worked On (circle)
LF /RF / LH / RH / ALL

Next Visit Due ___________

Hoof Condition / Notes ______________________________

____

____

____

____

____

____

Farrier / Hoof Care Record

Animal Name / ID ______________________________

Date ____________________

Farrier ____________________

Service Type (circle)
Trim / Full Set / Front Shoes / Reset / Corrective

Hoof Worked On (circle)
LF /RF / LH / RH / ALL

Next Visit Due ___________

Hoof Condition / Notes ______________________________

____

____

____

____

____

____

Farrier / Hoof Care Record

Animal Name / ID ________________________________

Date ____________________

Farrier ___________________

Service Type (circle)
Trim / Full Set / Front Shoes / Reset / Corrective

Hoof Worked On (circle)
LF /RF / LH / RH / ALL

Next Visit Due ___________

Hoof Condition / Notes ______________________________

____

____

____

____

____

____

Farrier / Hoof Care Record

Animal Name / ID ______________________________

Date ____________________

Farrier ____________________

Service Type (circle)
Trim / Full Set / Front Shoes / Reset / Corrective

Hoof Worked On (circle)
LF /RF / LH / RH / ALL

Next Visit Due ___________

Hoof Condition / Notes ______________________________

____

____

____

____

____

____

Farrier / Hoof Care Record

Animal Name / ID ________________________________
Date ____________________
Farrier ____________________

Service Type (circle)
Trim / Full Set / Front Shoes / Reset / Corrective

Hoof Worked On (circle)
LF /RF / LH / RH / ALL

Next Visit Due ___________

Hoof Condition / Notes ______________________________
____
____
____
____
____
____

Farrier / Hoof Care Record

Animal Name / ID ________________________________

Date ____________________

Farrier __________________

Service Type (circle)

Trim / Full Set / Front Shoes / Reset / Corrective

Hoof Worked On (circle)

LF /RF / LH / RH / ALL

Next Visit Due ___________

Hoof Condition / Notes ______________________________

____

____

____

____

____

____

Farrier / Hoof Care Record

Animal Name / ID ________________________________

Date ____________________

Farrier __________________

Service Type (circle)
Trim / Full Set / Front Shoes / Reset / Corrective

Hoof Worked On (circle)
LF /RF / LH / RH / ALL

Next Visit Due ___________

Hoof Condition / Notes ______________________________

____

____

____

____

____

____

Farrier / Hoof Care Record

Animal Name / ID ______________________________

Date ___________________

Farrier _________________

Service Type (circle)
Trim / Full Set / Front Shoes / Reset / Corrective

Hoof Worked On (circle)
LF /RF / LH / RH / ALL

Next Visit Due ___________

Hoof Condition / Notes _____________________________

____

____

____

____

____

____

Farrier / Hoof Care Record

Animal Name / ID ______________________________

Date ____________________

Farrier __________________

Service Type (circle)
Trim / Full Set / Front Shoes / Reset / Corrective

Hoof Worked On (circle)
LF /RF / LH / RH / ALL

Next Visit Due ___________

Hoof Condition / Notes ______________________________

____

____

____

____

____

____

Farrier / Hoof Care Record

Animal Name / ID ______________________________

Date ____________________

Farrier ___________________

Service Type (circle)
Trim / Full Set / Front Shoes / Reset / Corrective

Hoof Worked On (circle)
LF /RF / LH / RH / ALL

Next Visit Due ___________

Hoof Condition / Notes ______________________________
____
____
____
____
____
____

Farrier / Hoof Care Record

Animal Name / ID ________________________________

Date ____________________

Farrier __________________

Service Type (circle)
Trim / Full Set / Front Shoes / Reset / Corrective

Hoof Worked On (circle)
LF /RF / LH / RH / ALL

Next Visit Due ___________

Hoof Condition / Notes ______________________________

____

____

____

____

____

____

Farrier / Hoof Care Record

Animal Name / ID ________________________________

Date ____________________

Farrier ___________________

Service Type (circle)

Trim / Full Set / Front Shoes / Reset / Corrective

Hoof Worked On (circle)

LF /RF / LH / RH / ALL

Next Visit Due ___________

Hoof Condition / Notes ______________________________

____

____

____

____

____

____

Farrier / Hoof Care Record

Animal Name / ID ______________________________

Date ____________________

Farrier __________________

Service Type (circle)
Trim / Full Set / Front Shoes / Reset / Corrective

Hoof Worked On (circle)
LF /RF / LH / RH / ALL

Next Visit Due ___________

Hoof Condition / Notes ______________________________

____

____

____

____

____

____

Farrier / Hoof Care Record

Animal Name / ID ______________________________

Date ____________________

Farrier __________________

Service Type (circle)
Trim / Full Set / Front Shoes / Reset / Corrective

Hoof Worked On (circle)
LF /RF / LH / RH / ALL

Next Visit Due ___________

Hoof Condition / Notes ______________________________

____

____

____

____

____

____

Farrier / Hoof Care Record

Animal Name / ID ______________________________

Date ____________________

Farrier __________________

Service Type (circle)
Trim / Full Set / Front Shoes / Reset / Corrective

Hoof Worked On (circle)
LF /RF / LH / RH / ALL

Next Visit Due ___________

Hoof Condition / Notes ______________________________

____

____

____

____

____

____

Farrier / Hoof Care Record

Animal Name / ID ______________________________
Date ____________________
Farrier ___________________

Service Type (circle)
Trim / Full Set / Front Shoes / Reset / Corrective

Hoof Worked On (circle)
LF /RF / LH / RH / ALL

Next Visit Due ___________

Hoof Condition / Notes ______________________________

____

____

____

____

____

____

Farrier / Hoof Care Record

Animal Name / ID ________________________________

Date ____________________

Farrier __________________

Service Type (circle)
Trim / Full Set / Front Shoes / Reset / Corrective

Hoof Worked On (circle)
LF /RF / LH / RH / ALL

Next Visit Due ___________

Hoof Condition / Notes ______________________________

____

____

____

____

____

____

Farrier / Hoof Care Record

Animal Name / ID ________________________________

Date ____________________

Farrier __________________

Service Type (circle)

Trim / Full Set / Front Shoes / Reset / Corrective

Hoof Worked On (circle)

LF /RF / LH / RH / ALL

Next Visit Due ___________

Hoof Condition / Notes ______________________________

____

____

____

____

____

____

Farrier / Hoof Care Record

Animal Name / ID ________________________________
Date ____________________
Farrier __________________

Service Type (circle)
Trim / Full Set / Front Shoes / Reset / Corrective

Hoof Worked On (circle)
LF /RF / LH / RH / ALL

Next Visit Due ___________

Hoof Condition / Notes ______________________________

____

____

____

____

____

____

Farrier / Hoof Care Record

Animal Name / ID ______________________________

Date __________________

Farrier __________________

Service Type (circle)

Trim / Full Set / Front Shoes / Reset / Corrective

Hoof Worked On (circle)

LF /RF / LH / RH / ALL

Next Visit Due ___________

Hoof Condition / Notes ______________________________

____

____

____

____

____

____

Farrier / Hoof Care Record

Animal Name / ID ______________________________

Date ____________________

Farrier ___________________

Service Type (circle)
Trim / Full Set / Front Shoes / Reset / Corrective

Hoof Worked On (circle)
LF /RF / LH / RH / ALL

Next Visit Due ___________

Hoof Condition / Notes ______________________________

____

____

____

____

____

____

Farrier / Hoof Care Record

Animal Name / ID ________________________________

Date ____________________

Farrier ___________________

Service Type (circle)
Trim / Full Set / Front Shoes / Reset / Corrective

Hoof Worked On (circle)
LF /RF / LH / RH / ALL

Next Visit Due ___________

Hoof Condition / Notes ______________________________

____

____

____

____

____

____

Farrier / Hoof Care Record

Animal Name / ID ________________________________

Date ____________________

Farrier __________________

Service Type (circle)
Trim / Full Set / Front Shoes / Reset / Corrective

Hoof Worked On (circle)
LF /RF / LH / RH / ALL

Next Visit Due ___________

Hoof Condition / Notes ______________________________

____

____

____

____

____

____

Farrier / Hoof Care Record

Animal Name / ID ______________________________

Date ___________________

Farrier _________________

Service Type (circle)
Trim / Full Set / Front Shoes / Reset / Corrective

Hoof Worked On (circle)
LF /RF / LH / RH / ALL

Next Visit Due __________

Hoof Condition / Notes ____________________________

____

____

____

____

____

____

Farrier / Hoof Care Record

Animal Name / ID ________________________________

Date ____________________

Farrier __________________

Service Type (circle)
Trim / Full Set / Front Shoes / Reset / Corrective

Hoof Worked On (circle)
LF /RF / LH / RH / ALL

Next Visit Due ___________

Hoof Condition / Notes ______________________________

____

____

____

____

____

____

Farrier / Hoof Care Record

Animal Name / ID ______________________________

Date ____________________

Farrier ____________________

Service Type (circle)

Trim / Full Set / Front Shoes / Reset / Corrective

Hoof Worked On (circle)

LF /RF / LH / RH / ALL

Next Visit Due ___________

Hoof Condition / Notes ____________________________

____

____

____

____

____

____

Veterinary Medical Record

Animal Name / ID ______________________
Date of Visit ____________________________
Veterinarian Name / Clinic _______________

Reason for Visit ____________________________________
____
____
____
____

Diagnosis ____
____

Treatment ____
____
____
____
____
____
____
____

Follow- Up Date ______________________

Additional Notes____________________________________
____

Veterinary Medical Record

Animal Name / ID ______________________

Date of Visit __________________________

Veterinarian Name / Clinic _______________

Reason for Visit ____________________________________

____

____

____

____

Diagnosis ____

____

Treatment ____

____

____

____

____

____

____

____

Follow- Up Date ______________________

Additional Notes____________________________________

____

Veterinary Medical Record

Animal Name / ID ______________________________
Date of Visit ____________________________________
Veterinarian Name / Clinic _____________________

Reason for Visit ____
____
____
____
____

Diagnosis ____
____

Treatment ____
____
____
____
____
____
____
____

Follow- Up Date ______________________________

Additional Notes____
____

Veterinary Medical Record

Animal Name / ID ______________________

Date of Visit ___________________________

Veterinarian Name / Clinic _______________

Reason for Visit ______________________________________

____

____

____

____

Diagnosis ____

____

Treatment ____

____

____

____

____

____

____

____

Follow- Up Date _______________________

Additional Notes_____________________________________

____

Veterinary Medical Record

Animal Name / ID ___________________________

Date of Visit _________________________________

Veterinarian Name / Clinic ___________________

Reason for Visit ____

____

____

____

____

Diagnosis ____

____

Treatment ____

____

____

____

____

____

____

____

Follow- Up Date ___________________________

Additional Notes____

____

Veterinary Medical Record

Animal Name / ID ______________________

Date of Visit ______________________

Veterinarian Name / Clinic ______________________

Reason for Visit ______________________

____

____

____

____

Diagnosis ______________________

____

Treatment ______________________

____

____

____

____

____

____

____

Follow- Up Date ______________________

Additional Notes ______________________

____

Veterinary Medical Record

Animal Name / ID ____________________________

Date of Visit ________________________________

Veterinarian Name / Clinic ____________________

Reason for Visit ____

____

____

____

____

Diagnosis ____

____

Treatment ____

____

____

____

____

____

____

____

Follow- Up Date ____________________________

Additional Notes ____

____

Veterinary Medical Record

Animal Name / ID ______________________

Date of Visit ___________________________

Veterinarian Name / Clinic _______________

Reason for Visit ____________________________________

____

____

____

____

Diagnosis ____

____

Treatment _______________________________________

____

____

____

____

____

____

____

Follow- Up Date ______________________

Additional Notes__________________________________

____

Veterinary Medical Record

Animal Name / ID ______________________

Date of Visit ____________________________

Veterinarian Name / Clinic ________________

Reason for Visit ____________________________________

____

____

____

____

Diagnosis ____

____

Treatment ____

____

____

____

____

____

____

____

Follow- Up Date ______________________

Additional Notes___________________________________

____

Veterinary Medical Record

Animal Name / ID ____________________________

Date of Visit __________________________________

Veterinarian Name / Clinic ___________________

Reason for Visit ____

____

____

____

____

Diagnosis ____

____

Treatment ____

____

____

____

____

____

____

____

Follow- Up Date ____________________________

Additional Notes____

____

Veterinary Medical Record

Animal Name / ID ____________________________

Date of Visit ___________________________________

Veterinarian Name / Clinic ___________________

Reason for Visit ____

____

____

____

____

Diagnosis ____

____

Treatment ____

____

____

____

____

____

____

____

Follow- Up Date ____________________________

Additional Notes____

____

Veterinary Medical Record

Animal Name / ID ____________________

Date of Visit ____________________

Veterinarian Name / Clinic ____________________

Reason for Visit ____________________

____

____

____

____

Diagnosis ____________________

____

Treatment ____________________

____

____

____

____

____

____

____

Follow- Up Date ____________________

Additional Notes ____________________

____

Veterinary Medical Record

Animal Name / ID ____________________________

Date of Visit ________________________________

Veterinarian Name / Clinic ____________________

Reason for Visit ____

____

____

____

____

Diagnosis ____

____

Treatment ____

____

____

____

____

____

____

____

Follow- Up Date ___________________________

Additional Notes____

____

Veterinary Medical Record

Animal Name / ID ____________________________

Date of Visit ___________________________________

Veterinarian Name / Clinic ___________________

Reason for Visit ____

____

____

____

____

Diagnosis ____

____

Treatment ____

____

____

____

____

____

____

____

Follow- Up Date ____________________________

Additional Notes____

____

Veterinary Medical Record

Animal Name / ID ____________________________
Date of Visit __________________________________
Veterinarian Name / Clinic ____________________

Reason for Visit ____
____
____
____
____

Diagnosis ____
____

Treatment ____
____
____
____
____
____
____
____

Follow- Up Date ____________________________

Additional Notes____
____

Veterinary Medical Record

Animal Name / ID ______________________

Date of Visit ______________________

Veterinarian Name / Clinic ______________________

Reason for Visit ______________________

____

____

____

____

Diagnosis ______________________

____

Treatment ______________________

____

____

____

____

____

____

____

Follow- Up Date ______________________

Additional Notes ______________________

____

Veterinary Medical Record

Animal Name / ID ______________________

Date of Visit ____________________________

Veterinarian Name / Clinic _______________

Reason for Visit ____________________________________

____

____

____

____

Diagnosis ____

____

Treatment ____

____

____

____

____

____

____

____

Follow- Up Date _______________________

Additional Notes_____________________________________

____

Veterinary Medical Record

Animal Name / ID ______________________

Date of Visit ___________________________

Veterinarian Name / Clinic _______________

Reason for Visit ____________________________________

____

____

____

____

Diagnosis ____

____

Treatment ____

____

____

____

____

____

____

____

Follow- Up Date ______________________

Additional Notes____________________________________

____

Veterinary Medical Record

Animal Name / ID ___________________________

Date of Visit ________________________________

Veterinarian Name / Clinic ___________________

Reason for Visit ____

____

____

____

____

Diagnosis ____

____

Treatment ____

____

____

____

____

____

____

____

Follow- Up Date ___________________________

Additional Notes____

____

Veterinary Medical Record

Animal Name / ID ____________________________

Date of Visit __________________________________

Veterinarian Name / Clinic ____________________

Reason for Visit ____

____

____

____

____

Diagnosis ____

____

Treatment ____

____

____

____

____

____

____

____

Follow- Up Date ____________________________

Additional Notes____

____

Veterinary Medical Record

Animal Name / ID ______________________________

Date of Visit ____________________________________

Veterinarian Name / Clinic _____________________

Reason for Visit ____

____

____

____

____

Diagnosis ____

____

Treatment ____

____

____

____

____

____

____

____

Follow- Up Date ______________________________

Additional Notes ____

____

Veterinary Medical Record

Animal Name / ID ______________________

Date of Visit ____________________________

Veterinarian Name / Clinic ________________

Reason for Visit ____________________________________

____

____

____

____

Diagnosis ____

____

Treatment ____

____

____

____

____

____

____

____

Follow- Up Date ______________________

Additional Notes_____________________________________

____

Veterinary Medical Record

Animal Name / ID ______________________________

Date of Visit ____________________________________

Veterinarian Name / Clinic _____________________

Reason for Visit ____

____

____

____

____

Diagnosis ____

____

Treatment ____

____

____

____

____

____

____

____

Follow- Up Date ______________________________

Additional Notes____

____

Veterinary Medical Record

Animal Name / ID ____________________________

Date of Visit ____________________________________

Veterinarian Name / Clinic ___________________

Reason for Visit ____

____

____

____

____

Diagnosis ____

____

Treatment ____

____

____

____

____

____

____

____

Follow- Up Date ____________________________

Additional Notes____

____

Veterinary Medical Record

Animal Name / ID ______________________________

Date of Visit ____________________________________

Veterinarian Name / Clinic _____________________

Reason for Visit ____

____

____

____

____

Diagnosis ____

____

Treatment ____

____

____

____

____

____

____

____

Follow- Up Date ___________________________

Additional Notes____

____

Veterinary Medical Record

Animal Name / ID ____________________________

Date of Visit __________________________________

Veterinarian Name / Clinic ___________________

Reason for Visit ____

____

____

____

____

Diagnosis ____

____

Treatment ____

____

____

____

____

____

____

____

Follow- Up Date ____________________________

Additional Notes____

____

Veterinary Medical Record

Animal Name / ID ____________________________
Date of Visit __________________________________
Veterinarian Name / Clinic ___________________

Reason for Visit ____
____
____
____
____

Diagnosis ____
____

Treatment ____
____
____
____
____
____
____
____

Follow- Up Date ____________________________

Additional Notes____
____

Veterinary Medical Record

Animal Name / ID ___________________________
Date of Visit _________________________________
Veterinarian Name / Clinic __________________

Reason for Visit ____
____
____
____
____

Diagnosis ____
____

Treatment ____
____
____
____
____
____
____
____

Follow- Up Date ___________________________

Additional Notes____
____

Veterinary Medical Record

Animal Name / ID ______________________________

Date of Visit ____________________________________

Veterinarian Name / Clinic _____________________

Reason for Visit ____

____

____

____

____

Diagnosis ____

____

Treatment ____

____

____

____

____

____

____

____

Follow- Up Date ______________________________

Additional Notes ____

____

Veterinary Medical Record

Animal Name / ID ______________________

Date of Visit ___________________________

Veterinarian Name / Clinic ________________

Reason for Visit ______________________________________

____

____

____

____

Diagnosis ____

____

Treatment ____

____

____

____

____

____

____

____

Follow- Up Date ______________________

Additional Notes_____________________________________

____

Veterinary Medical Record

Animal Name / ID ______________________________

Date of Visit ____________________________________

Veterinarian Name / Clinic _____________________

Reason for Visit ____

____

____

____

____

Diagnosis ____

____

Treatment ____

____

____

____

____

____

____

____

Follow- Up Date ______________________________

Additional Notes____

____

Veterinary Medical Record

Animal Name / ID ________________________

Date of Visit ____________________________

Veterinarian Name / Clinic ________________

Reason for Visit ____

____

____

____

____

Diagnosis ____

____

Treatment ____

____

____

____

____

____

____

____

Follow- Up Date ________________________

Additional Notes ____

____

Veterinary Medical Record

Animal Name / ID ______________________

Date of Visit ______________________

Veterinarian Name / Clinic ______________________

Reason for Visit ______________________

____

____

____

____

Diagnosis ______________________

____

Treatment ______________________

____

____

____

____

____

____

____

Follow- Up Date ______________________

Additional Notes______________________

____

Veterinary Medical Record

Animal Name / ID ______________________

Date of Visit ___________________________

Veterinarian Name / Clinic _______________

Reason for Visit ____________________________________

____

____

____

____

Diagnosis ____

____

Treatment ____

____

____

____

____

____

____

____

Follow- Up Date ______________________

Additional Notes____________________________________

____

Veterinary Medical Record

Animal Name / ID ______________________________

Date of Visit ____________________________________

Veterinarian Name / Clinic _____________________

Reason for Visit ____

____

____

____

____

Diagnosis ____

____

Treatment ____

____

____

____

____

____

____

____

Follow- Up Date ______________________________

Additional Notes____

____

Veterinary Medical Record

Animal Name / ID ___________________________

Date of Visit _________________________________

Veterinarian Name / Clinic ___________________

Reason for Visit ____

____

____

____

____

Diagnosis ____

____

Treatment ____

____

____

____

____

____

____

____

Follow- Up Date ___________________________

Additional Notes____

____

Veterinary Medical Record

Animal Name / ID ______________________

Date of Visit ___________________________

Veterinarian Name / Clinic _______________

Reason for Visit ___________________________________

____

____

____

____

Diagnosis ____

____

Treatment _______________________________________

____

____

____

____

____

____

____

Follow- Up Date ______________________

Additional Notes___________________________________

____

Veterinary Medical Record

Animal Name / ID ______________________

Date of Visit ____________________________

Veterinarian Name / Clinic _______________

Reason for Visit ____________________________________

____

____

____

____

Diagnosis ____

____

Treatment ____

____

____

____

____

____

____

____

Follow- Up Date ______________________

Additional Notes___________________________________

____

Veterinary Medical Record

Animal Name / ID ______________________

Date of Visit ___________________________

Veterinarian Name / Clinic _______________

Reason for Visit _____________________________________

____

____

____

____

Diagnosis ____

____

Treatment ____

____

____

____

____

____

____

____

Follow- Up Date ______________________

Additional Notes_____________________________________

____

Veterinary Medical Record

Animal Name / ID ______________________

Date of Visit ___________________________

Veterinarian Name / Clinic ________________

Reason for Visit ____________________________________

____

____

____

____

Diagnosis ____

____

Treatment ____

____

____

____

____

____

____

____

Follow- Up Date ______________________

Additional Notes ____________________________________

____

Veterinary Medical Record

Animal Name / ID ______________________

Date of Visit __________________________

Veterinarian Name / Clinic _______________

Reason for Visit ____________________________________

____

____

____

____

Diagnosis ____

____

Treatment _______________________________________

____

____

____

____

____

____

____

Follow- Up Date ______________________

Additional Notes____________________________________

____

Veterinary Medical Record

Animal Name / ID ____________________________

Date of Visit ____________________________

Veterinarian Name / Clinic ____________________

Reason for Visit ____

____

____

____

____

Diagnosis ____

____

Treatment ____

____

____

____

____

____

____

____

Follow- Up Date ____________________________

Additional Notes ____

____

Veterinary Medical Record

Animal Name / ID ______________________

Date of Visit __________________________

Veterinarian Name / Clinic ________________

Reason for Visit ____________________________________

____

____

____

____

Diagnosis ____

____

Treatment ____

____

____

____

____

____

____

____

Follow- Up Date ______________________

Additional Notes_____________________________________

____

Veterinary Medical Record

Animal Name / ID ______________________________

Date of Visit ____________________________________

Veterinarian Name / Clinic _____________________

Reason for Visit ____

____

____

____

____

Diagnosis ____

____

Treatment ____

____

____

____

____

____

____

____

Follow- Up Date ______________________________

Additional Notes____

____

Veterinary Medical Record

Animal Name / ID ___________________________

Date of Visit _________________________________

Veterinarian Name / Clinic ___________________

Reason for Visit ____

____

____

____

____

Diagnosis ____

____

Treatment ____

____

____

____

____

____

____

____

Follow- Up Date ___________________________

Additional Notes____

____

Veterinary Medical Record

Animal Name / ID ______________________

Date of Visit ___________________________

Veterinarian Name / Clinic ________________

Reason for Visit ______________________________________

____

____

____

____

Diagnosis ____

____

Treatment ____

____

____

____

____

____

____

____

Follow- Up Date ______________________

Additional Notes_____________________________________

____

Veterinary Medical Record

Animal Name / ID ______________________

Date of Visit ______________________

Veterinarian Name / Clinic ________________

Reason for Visit ______________________________________

____

____

____

____

Diagnosis ____

____

Treatment ____

____

____

____

____

____

____

____

Follow- Up Date ______________________

Additional Notes______________________________________

____

Veterinary Medical Record

Animal Name / ID ___________________________

Date of Visit _________________________________

Veterinarian Name / Clinic __________________

Reason for Visit ____

____

____

____

____

Diagnosis ____

____

Treatment ____

____

____

____

____

____

____

____

Follow- Up Date ___________________________

Additional Notes ____

____

Veterinary Medical Record

Animal Name / ID ______________________

Date of Visit ___________________________

Veterinarian Name / Clinic ________________

Reason for Visit ______________________________________

____

____

____

____

Diagnosis ____

____

Treatment ____

____

____

____

____

____

____

____

Follow- Up Date ______________________

Additional Notes_____________________________________

____

Veterinary Medical Record

Animal Name / ID ____________________________

Date of Visit __________________________________

Veterinarian Name / Clinic ___________________

Reason for Visit ____

____

____

____

____

Diagnosis ____

____

Treatment ____

____

____

____

____

____

____

____

Follow- Up Date ____________________________

Additional Notes____

____

Breeding Record

Breeding Event / Attempt # ____________________
Dam (Mother) Animal Name / ID__________________________
Sire (Dad) Animal Name / ID _____________________________
Breeding Date ________________
Expected Due Date ___________
Birth Date ___________________
Number of Offspring _________

Offspring Details (Name / Sex / Color / Weight / Notes)

____

____

____

____

Complications / Notes

____

____

____

____

Veterinary Notes / Follow- Up

____

____

____

____

____

Breeding Record

Breeding Event / Attempt # ________________
Dam (Mother) Animal Name / ID______________________
Sire (Dad) Animal Name / ID _________________________
Breeding Date ______________
Expected Due Date __________
Birth Date _________________
Number of Offspring ________

Offspring Details (Name / Sex / Color / Weight / Notes)

____

____

____

____

Complications / Notes

____

____

____

____

Veterinary Notes / Follow- Up

____

____

____

____

____

Breeding Record

Breeding Event / Attempt # ________________
Dam (Mother) Animal Name / ID____________________
Sire (Dad) Animal Name / ID _______________________
Breeding Date _____________
Expected Due Date _________
Birth Date ________________
Number of Offspring _______

Offspring Details (Name / Sex / Color / Weight / Notes)

____

____

____

____

Complications / Notes

____

____

____

____

Veterinary Notes / Follow- Up

____

____

____

____

____

Breeding Record

Breeding Event / Attempt # ____________________
Dam (Mother) Animal Name / ID_________________________
Sire (Dad) Animal Name / ID ____________________________
Breeding Date ________________
Expected Due Date ___________
Birth Date ____________________
Number of Offspring _________

Offspring Details (Name / Sex / Color / Weight / Notes)

____

____

____

____

Complications / Notes

____

____

____

____

Veterinary Notes / Follow- Up

____

____

____

____

____

Breeding Record

Breeding Event / Attempt # ________________
Dam (Mother) Animal Name / ID______________________
Sire (Dad) Animal Name / ID _________________________
Breeding Date ______________
Expected Due Date __________
Birth Date _________________
Number of Offspring ________

Offspring Details (Name / Sex / Color / Weight / Notes)

____

____

____

____

Complications / Notes

____

____

____

____

Veterinary Notes / Follow- Up

____

____

____

____

____

Breeding Record

Breeding Event / Attempt # ________________
Dam (Mother) Animal Name / ID____________________
Sire (Dad) Animal Name / ID ______________________
Breeding Date _____________
Expected Due Date _________
Birth Date ________________
Number of Offspring _______

Offspring Details (Name / Sex / Color / Weight / Notes)

____

____

____

____

Complications / Notes

____

____

____

____

Veterinary Notes / Follow- Up

____

____

____

____

____

Breeding Record

Breeding Event / Attempt # ____________________
Dam (Mother) Animal Name / ID__________________________
Sire (Dad) Animal Name / ID _____________________________
Breeding Date ________________
Expected Due Date ___________
Birth Date ___________________
Number of Offspring _________

Offspring Details (Name / Sex / Color / Weight / Notes)

____

____

____

____

Complications / Notes

____

____

____

____

Veterinary Notes / Follow- Up

____

____

____

____

____

Breeding Record

Breeding Event / Attempt # ________________
Dam (Mother) Animal Name / ID____________________
Sire (Dad) Animal Name / ID ______________________
Breeding Date ______________
Expected Due Date __________
Birth Date ________________
Number of Offspring ________

Offspring Details (Name / Sex / Color / Weight / Notes)

____

____

____

____

Complications / Notes

____

____

____

____

Veterinary Notes / Follow- Up

____

____

____

____

____

Breeding Record

Breeding Event / Attempt # ____________________
Dam (Mother) Animal Name / ID_________________________
Sire (Dad) Animal Name / ID ____________________________
Breeding Date _______________
Expected Due Date ___________
Birth Date ___________________
Number of Offspring _________

Offspring Details (Name / Sex / Color / Weight / Notes)

____
____
____
____

Complications / Notes

____
____
____
____

Veterinary Notes / Follow- Up

____
____
____
____
____

Breeding Record

Breeding Event / Attempt # ___________________
Dam (Mother) Animal Name / ID________________________
Sire (Dad) Animal Name / ID ____________________________
Breeding Date ______________
Expected Due Date ___________
Birth Date ___________________
Number of Offspring ________

Offspring Details (Name / Sex / Color / Weight / Notes)

____

____

____

____

Complications / Notes

____

____

____

____

Veterinary Notes / Follow- Up

____

____

____

____

____

Breeding Record

Breeding Event / Attempt # ________________
Dam (Mother) Animal Name / ID_____________________
Sire (Dad) Animal Name / ID ________________________
Breeding Date ______________
Expected Due Date _________
Birth Date ________________
Number of Offspring _______

Offspring Details (Name / Sex / Color / Weight / Notes)

____

____

____

____

Complications / Notes

____

____

____

____

Veterinary Notes / Follow- Up

____

____

____

____

____

Breeding Record

Breeding Event / Attempt # ________________
Dam (Mother) Animal Name / ID____________________
Sire (Dad) Animal Name / ID ______________________
Breeding Date ______________
Expected Due Date _________
Birth Date _______________
Number of Offspring _______

Offspring Details (Name / Sex / Color / Weight / Notes)

____

____

____

____

Complications / Notes

____

____

____

____

Veterinary Notes / Follow- Up

____

____

____

____

____

Breeding Record

Breeding Event / Attempt # ________________
Dam (Mother) Animal Name / ID_____________________
Sire (Dad) Animal Name / ID _______________________
Breeding Date _____________
Expected Due Date _________
Birth Date ________________
Number of Offspring _______

Offspring Details (Name / Sex / Color / Weight / Notes)

____

____

____

____

Complications / Notes

____

____

____

____

Veterinary Notes / Follow- Up

____

____

____

____

____

Breeding Record

Breeding Event / Attempt # ____________________
Dam (Mother) Animal Name / ID_________________________
Sire (Dad) Animal Name / ID ____________________________
Breeding Date ________________
Expected Due Date ___________
Birth Date ___________________
Number of Offspring _________

Offspring Details (Name / Sex / Color / Weight / Notes)

____

____

____

____

Complications / Notes

____

____

____

____

Veterinary Notes / Follow- Up

____

____

____

____

____

Breeding Record

Breeding Event / Attempt # ________________
Dam (Mother) Animal Name / ID______________________
Sire (Dad) Animal Name / ID ________________________
Breeding Date ______________
Expected Due Date __________
Birth Date _________________
Number of Offspring ________

Offspring Details (Name / Sex / Color / Weight / Notes)

____

____

____

____

Complications / Notes

____

____

____

____

Veterinary Notes / Follow- Up

____

____

____

____

____

Breeding Record

Breeding Event / Attempt # ___________________
Dam (Mother) Animal Name / ID_________________________
Sire (Dad) Animal Name / ID ____________________________
Breeding Date ________________
Expected Due Date ___________
Birth Date ___________________
Number of Offspring _________

Offspring Details (Name / Sex / Color / Weight / Notes)

____

____

____

____

Complications / Notes

____

____

____

____

Veterinary Notes / Follow- Up

____

____

____

____

____

Breeding Record

Breeding Event / Attempt # ____________________
Dam (Mother) Animal Name / ID__________________________
Sire (Dad) Animal Name / ID _____________________________
Breeding Date ________________
Expected Due Date ___________
Birth Date ____________________
Number of Offspring _________

Offspring Details (Name / Sex / Color / Weight / Notes)

____

____

____

____

Complications / Notes

____

____

____

____

Veterinary Notes / Follow- Up

____

____

____

____

____

Breeding Record

Breeding Event / Attempt # ____________________
Dam (Mother) Animal Name / ID_________________________
Sire (Dad) Animal Name / ID ____________________________
Breeding Date ________________
Expected Due Date ___________
Birth Date ___________________
Number of Offspring _________

Offspring Details (Name / Sex / Color / Weight / Notes)

____

____

____

____

Complications / Notes

____

____

____

____

Veterinary Notes / Follow- Up

____

____

____

____

____

Breeding Record

Breeding Event / Attempt # ________________
Dam (Mother) Animal Name / ID______________________
Sire (Dad) Animal Name / ID ________________________
Breeding Date _____________
Expected Due Date _________
Birth Date ________________
Number of Offspring _______

Offspring Details (Name / Sex / Color / Weight / Notes)

____

____

____

____

Complications / Notes

____

____

____

____

Veterinary Notes / Follow- Up

____

____

____

____

____

Breeding Record

Breeding Event / Attempt # ________________
Dam (Mother) Animal Name / ID____________________
Sire (Dad) Animal Name / ID _______________________
Breeding Date ______________
Expected Due Date _________
Birth Date ________________
Number of Offspring _______

Offspring Details (Name / Sex / Color / Weight / Notes)

____

____

____

____

Complications / Notes

____

____

____

____

Veterinary Notes / Follow- Up

____

____

____

____

____

Breeding Record

Breeding Event / Attempt # ____________________
Dam (Mother) Animal Name / ID__________________________
Sire (Dad) Animal Name / ID _____________________________
Breeding Date ________________
Expected Due Date ___________
Birth Date ____________________
Number of Offspring _________

Offspring Details (Name / Sex / Color / Weight / Notes)

____

____

____

____

Complications / Notes

____

____

____

____

Veterinary Notes / Follow- Up

____

____

____

____

____

Breeding Record

Breeding Event / Attempt # ____________________
Dam (Mother) Animal Name / ID_________________________
Sire (Dad) Animal Name / ID ____________________________
Breeding Date ________________
Expected Due Date ___________
Birth Date ___________________
Number of Offspring _________

Offspring Details (Name / Sex / Color / Weight / Notes)

____
____
____
____

Complications / Notes

____
____
____
____

Veterinary Notes / Follow- Up

____
____
____
____
____

Breeding Record

Breeding Event / Attempt # ________________
Dam (Mother) Animal Name / ID_____________________
Sire (Dad) Animal Name / ID _______________________
Breeding Date _____________
Expected Due Date _________
Birth Date ________________
Number of Offspring _______

Offspring Details (Name / Sex / Color / Weight / Notes)

____

____

____

____

Complications / Notes

____

____

____

____

Veterinary Notes / Follow- Up

____

____

____

____

____

Breeding Record

Breeding Event / Attempt # __________________
Dam (Mother) Animal Name / ID_______________________
Sire (Dad) Animal Name / ID __________________________
Breeding Date _______________
Expected Due Date ___________
Birth Date __________________
Number of Offspring ________

Offspring Details (Name / Sex / Color / Weight / Notes)

____

____

____

____

Complications / Notes

____

____

____

____

Veterinary Notes / Follow- Up

____

____

____

____

____

Breeding Record

Breeding Event / Attempt # __________________
Dam (Mother) Animal Name / ID________________________
Sire (Dad) Animal Name / ID ___________________________
Breeding Date ______________
Expected Due Date __________
Birth Date __________________
Number of Offspring ________

Offspring Details (Name / Sex / Color / Weight / Notes)

____

____

____

____

Complications / Notes

____

____

____

____

Veterinary Notes / Follow- Up

____

____

____

____

____

Breeding Record

Breeding Event / Attempt # ________________
Dam (Mother) Animal Name / ID____________________
Sire (Dad) Animal Name / ID _______________________
Breeding Date _____________
Expected Due Date _________
Birth Date ________________
Number of Offspring _______

Offspring Details (Name / Sex / Color / Weight / Notes)

____
____
____
____

Complications / Notes

____
____
____
____

Veterinary Notes / Follow- Up

____
____
____
____
____

Behavioral / Observation Record

Animal Name / ID ________________________

Date ___________

Observed Behavior

____

____

____

____

Possible Cause / Trigger

____

____

____

Action Taken

____

____

____

Follow- Up Notes ______________________________________

____

____

____

____

____

Behavioral / Observation Record

Animal Name / ID ______________________________

Date _______________

Observed Behavior

____

____

____

____

Possible Cause / Trigger

____

____

____

Action Taken

____

____

____

Follow- Up Notes ____

____

____

____

____

Behavioral / Observation Record

Animal Name / ID ______________________

Date ___________

Observed Behavior

____

____

____

____

Possible Cause / Trigger

____

____

____

Action Taken

____

____

____

Follow- Up Notes ______________________________

____

____

____

____

Behavioral / Observation Record

Animal Name / ID ______________________

Date ___________

Observed Behavior

____

____

____

____

Possible Cause / Trigger

____

____

____

Action Taken

____

____

____

Follow- Up Notes ___________________________________

____

____

____

____

Behavioral / Observation Record

Animal Name / ID ________________________

Date ____________

Observed Behavior

____

____

____

____

Possible Cause / Trigger

____

____

____

Action Taken

____

____

____

Follow- Up Notes____

____

____

____

____

Behavioral / Observation Record

Animal Name / ID ____________________________

Date ______________

Observed Behavior

____

____

____

____

Possible Cause / Trigger

____

____

____

Action Taken

____

____

____

Follow- Up Notes____

____

____

____

____

Behavioral / Observation Record

Animal Name / ID ____________________________

Date ______________

Observed Behavior

____

____

____

____

Possible Cause / Trigger

____

____

____

Action Taken

____

____

____

Follow- Up Notes ____

____

____

____

____

Behavioral / Observation Record

Animal Name / ID ______________________

Date ___________

Observed Behavior

____

____

____

____

Possible Cause / Trigger

____

____

____

Action Taken

____

____

____

Follow- Up Notes ________________________________

____

____

____

____

Behavioral / Observation Record

Animal Name / ID ______________________

Date ___________

Observed Behavior

____

____

____

____

Possible Cause / Trigger

____

____

____

Action Taken

____

____

____

Follow- Up Notes ________________________________

____

____

____

____

Behavioral / Observation Record

Animal Name / ID ______________________

Date ___________

Observed Behavior

____

____

____

____

Possible Cause / Trigger

____

____

____

Action Taken

____

____

____

Follow- Up Notes ___________________________________

____

____

____

____

Behavioral / Observation Record

Animal Name / ID ______________________________

Date _______________

Observed Behavior

____

____

____

____

Possible Cause / Trigger

____

____

____

Action Taken

____

____

____

Follow- Up Notes ____

____

____

____

____

Behavioral / Observation Record

Animal Name / ID ____________________________

Date ______________

Observed Behavior

____

____

____

____

Possible Cause / Trigger

____

____

____

Action Taken

____

____

____

Follow- Up Notes

____

____

____

____

____

Behavioral / Observation Record

Animal Name / ID ______________________________
Date _______________

Observed Behavior

____

____

____

____

Possible Cause / Trigger

____

____

____

Action Taken

____

____

____

Follow- Up Notes ____

____

____

____

____

Behavioral / Observation Record

Animal Name / ID ______________________

Date ___________

Observed Behavior

____

____

____

____

Possible Cause / Trigger

____

____

____

Action Taken

____

____

____

Follow- Up Notes______________________________

____

____

____

____

Behavioral / Observation Record

Animal Name / ID ______________________

Date ___________

Observed Behavior

____

____

____

____

Possible Cause / Trigger

____

____

____

Action Taken

____

____

____

Follow- Up Notes ________________________________

____

____

____

____

Behavioral / Observation Record

Animal Name / ID ______________________

Date ___________

Observed Behavior

____

____

____

____

Possible Cause / Trigger

____

____

____

Action Taken

____

____

____

Follow- Up Notes ________________________________

____

____

____

____

Behavioral / Observation Record

Animal Name / ID ______________________
Date ___________

Observed Behavior

____
____
____
____

Possible Cause / Trigger

____
____
____

Action Taken

____
____
____

Follow- Up Notes___________________________________

____
____
____
____

Behavioral / Observation Record

Animal Name / ID ______________________

Date ___________

Observed Behavior

____

____

____

____

Possible Cause / Trigger

____

____

____

Action Taken

____

____

____

Follow- Up Notes ______________________________

____

____

____

____

Behavioral / Observation Record

Animal Name / ID ______________________________

Date ______________

Observed Behavior

____

____

____

____

Possible Cause / Trigger

____

____

____

Action Taken

____

____

____

Follow- Up Notes ____

____

____

____

____

Behavioral / Observation Record

Animal Name / ID ____________________________

Date ______________

Observed Behavior

____

____

____

____

Possible Cause / Trigger

____

____

____

Action Taken

____

____

____

Follow- Up Notes ____

____

____

____

____

Behavioral / Observation Record

Animal Name / ID ____________________________

Date ______________

Observed Behavior

____

____

____

____

Possible Cause / Trigger

____

____

____

Action Taken

____

____

____

Follow- Up Notes ____

____

____

____

____

Behavioral / Observation Record

Animal Name / ID ______________________

Date ___________

Observed Behavior

____

____

____

____

Possible Cause / Trigger

____

____

____

Action Taken

____

____

____

Follow- Up Notes ______________________________

____

____

____

____

Behavioral / Observation Record

Animal Name / ID ____________________________

Date ______________

Observed Behavior

____

____

____

____

Possible Cause / Trigger

____

____

____

Action Taken

____

____

____

Follow- Up Notes ____

____

____

____

____

Behavioral / Observation Record

Animal Name / ID ______________________

Date ___________

Observed Behavior

____

____

____

____

Possible Cause / Trigger

____

____

____

Action Taken

____

____

____

Follow- Up Notes __________________________________

____

____

____

____

Behavioral / Observation Record

Animal Name / ID ____________________________

Date ______________

Observed Behavior

____

____

____

____

Possible Cause / Trigger

____

____

____

Action Taken

____

____

____

Follow- Up Notes ____

____

____

____

____

Behavioral / Observation Record

Animal Name / ID ______________________

Date ___________

Observed Behavior

____

____

____

____

Possible Cause / Trigger

____

____

____

Action Taken

____

____

____

Follow- Up Notes___________________________________

____

____

____

____

Feeding Log

Animal Name / ID/ Group ______________________________

Location / Pen / Pasture ________________________________

Date ________________

Feeding Time (Circle) Morning / Midday / Evening / Other

Feed Type ____________________

Amount Given ________________

Supplements / Medications _____________________________

____

Fed By ______________

Additional Notes _______________________________________

____

____

Animal Name / ID/ Group ______________________________

Location / Pen / Pasture ________________________________

Date ________________

Feeding Time (Circle) Morning / Midday / Evening / Other

Feed Type ____________________

Amount Given ________________

Supplements / Medications _____________________________

____

Fed By ______________

Additional Notes _______________________________________

____

____

Feeding Log

Animal Name / ID/ Group ______________________________
Location / Pen / Pasture ________________________________
Date ________________
Feeding Time (Circle) Morning / Midday / Evening / Other
Feed Type ____________________
Amount Given ________________
Supplements / Medications _____________________________

____
Fed By ______________
Additional Notes _______________________________________
____
____

Animal Name / ID/ Group ______________________________
Location / Pen / Pasture ________________________________
Date ________________
Feeding Time (Circle) Morning / Midday / Evening / Other
Feed Type ____________________
Amount Given ________________
Supplements / Medications _____________________________

____
Fed By ______________
Additional Notes _______________________________________
____
____

Feeding Log

Animal Name / ID/ Group ____________________________

Location / Pen / Pasture ______________________________

Date _______________

Feeding Time (Circle) Morning / Midday / Evening / Other

Feed Type ___________________

Amount Given _______________

Supplements / Medications ___________________________

____

Fed By _____________

Additional Notes____________________________________

____

____

Animal Name / ID/ Group ____________________________

Location / Pen / Pasture ______________________________

Date _______________

Feeding Time (Circle) Morning / Midday / Evening / Other

Feed Type ___________________

Amount Given _______________

Supplements / Medications ___________________________

____

Fed By _____________

Additional Notes_____________________________________

____

____

Feeding Log

Animal Name / ID/ Group ______________________________
Location / Pen / Pasture ______________________________
Date ________________
Feeding Time (Circle) Morning / Midday / Evening / Other
Feed Type ____________________
Amount Given ________________
Supplements / Medications _____________________________
____
Fed By _____________
Additional Notes_______________________________________
____
____

Animal Name / ID/ Group ______________________________
Location / Pen / Pasture ______________________________
Date ________________
Feeding Time (Circle) Morning / Midday / Evening / Other
Feed Type ____________________
Amount Given ________________
Supplements / Medications _____________________________
____
Fed By _____________
Additional Notes_______________________________________
____
____

Feeding Log

Animal Name / ID/ Group ______________________________
Location / Pen / Pasture ________________________________
Date ________________
Feeding Time (Circle) Morning / Midday / Evening / Other
Feed Type ____________________
Amount Given ________________
Supplements / Medications _____________________________

____

Fed By ______________
Additional Notes _______________________________________

____

____

Animal Name / ID/ Group ______________________________
Location / Pen / Pasture ________________________________
Date ________________
Feeding Time (Circle) Morning / Midday / Evening / Other
Feed Type ____________________
Amount Given ________________
Supplements / Medications _____________________________

____

Fed By ______________
Additional Notes_______________________________________

____

____

Feeding Log

Animal Name / ID/ Group ______________________________
Location / Pen / Pasture ________________________________
Date ________________
Feeding Time (Circle) Morning / Midday / Evening / Other
Feed Type ____________________
Amount Given ________________
Supplements / Medications _____________________________

____

Fed By ______________
Additional Notes _______________________________________

____

____

Animal Name / ID/ Group ______________________________
Location / Pen / Pasture ________________________________
Date ________________
Feeding Time (Circle) Morning / Midday / Evening / Other
Feed Type ____________________
Amount Given ________________
Supplements / Medications _____________________________

____

Fed By ______________
Additional Notes _______________________________________

____

____

Feeding Log

Animal Name / ID/ Group ______________________________

Location / Pen / Pasture ________________________________

Date ________________

Feeding Time (Circle) Morning / Midday / Evening / Other

Feed Type ____________________

Amount Given ________________

Supplements / Medications _____________________________

____

Fed By ______________

Additional Notes _______________________________________

____

____

Animal Name / ID/ Group ______________________________

Location / Pen / Pasture ________________________________

Date ________________

Feeding Time (Circle) Morning / Midday / Evening / Other

Feed Type ____________________

Amount Given ________________

Supplements / Medications _____________________________

____

Fed By ______________

Additional Notes _______________________________________

____

____

Feeding Log

Animal Name / ID/ Group ______________________________
Location / Pen / Pasture ________________________________
Date ________________
Feeding Time (Circle) Morning / Midday / Evening / Other
Feed Type ____________________
Amount Given ________________
Supplements / Medications _____________________________
____
Fed By ______________
Additional Notes ______________________________________
____
____

Animal Name / ID/ Group ______________________________
Location / Pen / Pasture ________________________________
Date ________________
Feeding Time (Circle) Morning / Midday / Evening / Other
Feed Type ____________________
Amount Given ________________
Supplements / Medications _____________________________
____
Fed By ______________
Additional Notes ______________________________________
____
____

Feeding Log

Animal Name / ID/ Group ______________________________
Location / Pen / Pasture ________________________________
Date ________________
Feeding Time (Circle) Morning / Midday / Evening / Other
Feed Type ____________________
Amount Given ________________
Supplements / Medications _____________________________

____
Fed By ______________
Additional Notes _______________________________________

____

____

Animal Name / ID/ Group ______________________________
Location / Pen / Pasture ________________________________
Date ________________
Feeding Time (Circle) Morning / Midday / Evening / Other
Feed Type ____________________
Amount Given ________________
Supplements / Medications _____________________________

____
Fed By ______________
Additional Notes _______________________________________

____

____

Feeding Log

Animal Name / ID/ Group ______________________________
Location / Pen / Pasture ________________________________
Date ________________
Feeding Time (Circle) Morning / Midday / Evening / Other
Feed Type ____________________
Amount Given ________________
Supplements / Medications _____________________________

____
Fed By ______________
Additional Notes _____________________________________
____
____

Animal Name / ID/ Group ______________________________
Location / Pen / Pasture ________________________________
Date ________________
Feeding Time (Circle) Morning / Midday / Evening / Other
Feed Type ____________________
Amount Given ________________
Supplements / Medications _____________________________

____
Fed By ______________
Additional Notes _____________________________________
____
____

Feeding Log

Animal Name / ID/ Group ______________________________
Location / Pen / Pasture _________________________________
Date ________________
Feeding Time (Circle) Morning / Midday / Evening / Other
Feed Type ____________________
Amount Given ________________
Supplements / Medications _____________________________

____
Fed By ______________
Additional Notes _______________________________________

____

____

Animal Name / ID/ Group ______________________________
Location / Pen / Pasture _________________________________
Date ________________
Feeding Time (Circle) Morning / Midday / Evening / Other
Feed Type ____________________
Amount Given ________________
Supplements / Medications _____________________________

____
Fed By ______________
Additional Notes _______________________________________

____

____

Feeding Log

Animal Name / ID/ Group ______________________________
Location / Pen / Pasture ________________________________
Date ________________
Feeding Time (Circle) Morning / Midday / Evening / Other
Feed Type ____________________
Amount Given ________________
Supplements / Medications ____________________________

____

Fed By ______________
Additional Notes _______________________________________

____

____

Animal Name / ID/ Group ______________________________
Location / Pen / Pasture ________________________________
Date ________________
Feeding Time (Circle) Morning / Midday / Evening / Other
Feed Type ____________________
Amount Given ________________
Supplements / Medications ____________________________

____

Fed By ______________
Additional Notes _______________________________________

____

____

Feeding Log

Animal Name / ID/ Group ______________________________
Location / Pen / Pasture ________________________________
Date ________________
Feeding Time (Circle) Morning / Midday / Evening / Other
Feed Type ____________________
Amount Given ________________
Supplements / Medications _____________________________

____
Fed By ______________
Additional Notes _______________________________________

____

____

Animal Name / ID/ Group ______________________________
Location / Pen / Pasture ________________________________
Date ________________
Feeding Time (Circle) Morning / Midday / Evening / Other
Feed Type ____________________
Amount Given ________________
Supplements / Medications _____________________________

____
Fed By ______________
Additional Notes _______________________________________

____

____

Feeding Log

Animal Name / ID/ Group ________________________________
Location / Pen / Pasture ___________________________________
Date _________________
Feeding Time (Circle) Morning / Midday / Evening / Other
Feed Type ______________________
Amount Given _________________
Supplements / Medications _______________________________

____

Fed By _______________
Additional Notes ____

____

____

Animal Name / ID/ Group ________________________________
Location / Pen / Pasture ___________________________________
Date _________________
Feeding Time (Circle) Morning / Midday / Evening / Other
Feed Type ______________________
Amount Given _________________
Supplements / Medications _______________________________

____

Fed By _______________
Additional Notes ____

____

____

Feeding Log

Animal Name / ID/ Group ______________________________
Location / Pen / Pasture ________________________________
Date ________________
Feeding Time (Circle) Morning / Midday / Evening / Other
Feed Type ____________________
Amount Given ________________
Supplements / Medications _____________________________

____
Fed By ______________
Additional Notes _______________________________________

____

____

Animal Name / ID/ Group ______________________________
Location / Pen / Pasture ________________________________
Date ________________
Feeding Time (Circle) Morning / Midday / Evening / Other
Feed Type ____________________
Amount Given ________________
Supplements / Medications _____________________________

____
Fed By ______________
Additional Notes _______________________________________

____

____

Feeding Log

Animal Name / ID/ Group ______________________________

Location / Pen / Pasture ________________________________

Date ________________

Feeding Time (Circle) Morning / Midday / Evening / Other

Feed Type ____________________

Amount Given ________________

Supplements / Medications _____________________________

____

Fed By ______________

Additional Notes _____________________________________

____

____

Animal Name / ID/ Group ______________________________

Location / Pen / Pasture ________________________________

Date ________________

Feeding Time (Circle) Morning / Midday / Evening / Other

Feed Type ____________________

Amount Given ________________

Supplements / Medications _____________________________

____

Fed By ______________

Additional Notes _____________________________________

____

____

Feeding Log

Animal Name / ID/ Group ______________________________
Location / Pen / Pasture ________________________________
Date ________________
Feeding Time (Circle) Morning / Midday / Evening / Other
Feed Type ____________________
Amount Given ________________
Supplements / Medications _____________________________

____

Fed By ______________
Additional Notes _______________________________________

____

____

Animal Name / ID/ Group ______________________________
Location / Pen / Pasture ________________________________
Date ________________
Feeding Time (Circle) Morning / Midday / Evening / Other
Feed Type ____________________
Amount Given ________________
Supplements / Medications _____________________________

____

Fed By ______________
Additional Notes _______________________________________

____

____

Feeding Log

Animal Name / ID/ Group ____________________________
Location / Pen / Pasture ______________________________
Date _______________
Feeding Time (Circle) Morning / Midday / Evening / Other
Feed Type ___________________
Amount Given _______________
Supplements / Medications ___________________________

____

Fed By _____________
Additional Notes____________________________________

____

____

Animal Name / ID/ Group ____________________________
Location / Pen / Pasture ______________________________
Date _______________
Feeding Time (Circle) Morning / Midday / Evening / Other
Feed Type ___________________
Amount Given _______________
Supplements / Medications ___________________________

____

Fed By _____________
Additional Notes ____________________________________

____

____

Feeding Log

Animal Name / ID/ Group ______________________________
Location / Pen / Pasture ________________________________
Date ________________
Feeding Time (Circle) Morning / Midday / Evening / Other
Feed Type ____________________
Amount Given ________________
Supplements / Medications _____________________________

____

Fed By ______________
Additional Notes _______________________________________

____

____

Animal Name / ID/ Group ______________________________
Location / Pen / Pasture ________________________________
Date ________________
Feeding Time (Circle) Morning / Midday / Evening / Other
Feed Type ____________________
Amount Given ________________
Supplements / Medications _____________________________

____

Fed By ______________
Additional Notes _______________________________________

____

____

Feeding Log

Animal Name / ID/ Group ______________________________

Location / Pen / Pasture ________________________________

Date ________________

Feeding Time (Circle) Morning / Midday / Evening / Other

Feed Type ____________________

Amount Given ________________

Supplements / Medications _____________________________

____

Fed By ______________

Additional Notes ______________________________________

____

____

Animal Name / ID/ Group ______________________________

Location / Pen / Pasture ________________________________

Date ________________

Feeding Time (Circle) Morning / Midday / Evening / Other

Feed Type ____________________

Amount Given ________________

Supplements / Medications _____________________________

____

Fed By ______________

Additional Notes ______________________________________

____

____

Feeding Log

Animal Name / ID/ Group ______________________________
Location / Pen / Pasture ________________________________
Date ________________
Feeding Time (Circle) Morning / Midday / Evening / Other
Feed Type ____________________
Amount Given ________________
Supplements / Medications _____________________________

____
Fed By ______________
Additional Notes _______________________________________
____
____

Animal Name / ID/ Group ______________________________
Location / Pen / Pasture ________________________________
Date ________________
Feeding Time (Circle) Morning / Midday / Evening / Other
Feed Type ____________________
Amount Given ________________
Supplements / Medications _____________________________

____
Fed By ______________
Additional Notes _______________________________________
____
____

Feeding Log

Animal Name / ID/ Group ______________________________
Location / Pen / Pasture ________________________________
Date ________________
Feeding Time (Circle) Morning / Midday / Evening / Other
Feed Type ____________________
Amount Given ________________
Supplements / Medications _____________________________

____

Fed By ______________
Additional Notes ____

____

____

Animal Name / ID/ Group ______________________________
Location / Pen / Pasture ________________________________
Date ________________
Feeding Time (Circle) Morning / Midday / Evening / Other
Feed Type ____________________
Amount Given ________________
Supplements / Medications _____________________________

____

Fed By ______________
Additional Notes ____

____

____

Feeding Log

Animal Name / ID/ Group ______________________________
Location / Pen / Pasture ________________________________
Date ________________
Feeding Time (Circle) Morning / Midday / Evening / Other
Feed Type ____________________
Amount Given ________________
Supplements / Medications _____________________________

____
Fed By ______________
Additional Notes ______________________________________

____

____

Animal Name / ID/ Group ______________________________
Location / Pen / Pasture ________________________________
Date ________________
Feeding Time (Circle) Morning / Midday / Evening / Other
Feed Type ____________________
Amount Given ________________
Supplements / Medications _____________________________

____
Fed By ______________
Additional Notes ______________________________________

____

____

Feeding Log

Animal Name / ID/ Group ______________________________

Location / Pen / Pasture ________________________________

Date ________________

Feeding Time (Circle) Morning / Midday / Evening / Other

Feed Type ____________________

Amount Given ________________

Supplements / Medications ____________________________

____

Fed By _____________

Additional Notes _______________________________________

____

____

Animal Name / ID/ Group ______________________________

Location / Pen / Pasture ________________________________

Date ________________

Feeding Time (Circle) Morning / Midday / Evening / Other

Feed Type ____________________

Amount Given ________________

Supplements / Medications ____________________________

____

Fed By _____________

Additional Notes _______________________________________

____

____

Feeding Log

Animal Name / ID/ Group ______________________________
Location / Pen / Pasture ________________________________
Date ________________
Feeding Time (Circle) Morning / Midday / Evening / Other
Feed Type ____________________
Amount Given ________________
Supplements / Medications _____________________________

____

Fed By ______________
Additional Notes _______________________________________

____

____

Animal Name / ID/ Group ______________________________
Location / Pen / Pasture ________________________________
Date ________________
Feeding Time (Circle) Morning / Midday / Evening / Other
Feed Type ____________________
Amount Given ________________
Supplements / Medications _____________________________

____

Fed By ______________
Additional Notes _______________________________________

____

____

Feeding Log

Animal Name / ID/ Group ____________________________
Location / Pen / Pasture ______________________________
Date _______________
Feeding Time (Circle) Morning / Midday / Evening / Other
Feed Type ___________________
Amount Given _______________
Supplements / Medications ___________________________
____
Fed By _____________
Additional Notes ____________________________________
____
____

Animal Name / ID/ Group ____________________________
Location / Pen / Pasture ______________________________
Date _______________
Feeding Time (Circle) Morning / Midday / Evening / Other
Feed Type ___________________
Amount Given _______________
Supplements / Medications ___________________________
____
Fed By _____________
Additional Notes ____________________________________
____
____

Feeding Log

Animal Name / ID/ Group ______________________________
Location / Pen / Pasture ________________________________
Date ________________
Feeding Time (Circle) Morning / Midday / Evening / Other
Feed Type ____________________
Amount Given ________________
Supplements / Medications _____________________________

____

Fed By ______________
Additional Notes _______________________________________

____

____

Animal Name / ID/ Group ______________________________
Location / Pen / Pasture ________________________________
Date ________________
Feeding Time (Circle) Morning / Midday / Evening / Other
Feed Type ____________________
Amount Given ________________
Supplements / Medications _____________________________

____

Fed By ______________
Additional Notes _______________________________________

____

____

Feeding Log

Animal Name / ID/ Group ______________________
Location / Pen / Pasture ______________________
Date ____________
Feeding Time (Circle) Morning / Midday / Evening / Other
Feed Type ________________
Amount Given ____________
Supplements / Medications ______________________

____

Fed By ____________
Additional Notes ______________________________

____

____

Animal Name / ID/ Group ______________________
Location / Pen / Pasture ______________________
Date ____________
Feeding Time (Circle) Morning / Midday / Evening / Other
Feed Type ________________
Amount Given ____________
Supplements / Medications ______________________

____

Fed By ____________
Additional Notes ______________________________

____

____

Feeding Log

Animal Name / ID/ Group ______________________________
Location / Pen / Pasture _________________________________
Date ________________
Feeding Time (Circle) Morning / Midday / Evening / Other
Feed Type ____________________
Amount Given ________________
Supplements / Medications _____________________________

____
Fed By ______________
Additional Notes _______________________________________
____
____

Animal Name / ID/ Group ______________________________
Location / Pen / Pasture _________________________________
Date ________________
Feeding Time (Circle) Morning / Midday / Evening / Other
Feed Type ____________________
Amount Given ________________
Supplements / Medications _____________________________

____
Fed By ______________
Additional Notes _______________________________________
____
____

Feeding Log

Animal Name / ID/ Group ______________________________
Location / Pen / Pasture ________________________________
Date ________________
Feeding Time (Circle) Morning / Midday / Evening / Other
Feed Type ____________________
Amount Given ________________
Supplements / Medications _____________________________

____
Fed By ______________
Additional Notes _____________________________________
____
____

Animal Name / ID/ Group ______________________________
Location / Pen / Pasture ________________________________
Date ________________
Feeding Time (Circle) Morning / Midday / Evening / Other
Feed Type ____________________
Amount Given ________________
Supplements / Medications _____________________________

____
Fed By ______________
Additional Notes _____________________________________
____
____

Feeding Log

Animal Name / ID/ Group ______________________________
Location / Pen / Pasture ________________________________
Date ________________
Feeding Time (Circle) Morning / Midday / Evening / Other
Feed Type ____________________
Amount Given ________________
Supplements / Medications _____________________________

____
Fed By ______________
Additional Notes _______________________________________
____
____

Animal Name / ID/ Group ______________________________
Location / Pen / Pasture ________________________________
Date ________________
Feeding Time (Circle) Morning / Midday / Evening / Other
Feed Type ____________________
Amount Given ________________
Supplements / Medications _____________________________

____
Fed By ______________
Additional Notes _______________________________________
____
____

Feeding Log

Animal Name / ID/ Group ______________________________

Location / Pen / Pasture ________________________________

Date ________________

Feeding Time (Circle) Morning / Midday / Evening / Other

Feed Type ____________________

Amount Given ________________

Supplements / Medications _____________________________

____

Fed By ______________

Additional Notes _______________________________________

____

____

Animal Name / ID/ Group ______________________________

Location / Pen / Pasture ________________________________

Date ________________

Feeding Time (Circle) Morning / Midday / Evening / Other

Feed Type ____________________

Amount Given ________________

Supplements / Medications _____________________________

____

Fed By ______________

Additional Notes _______________________________________

____

____

Feeding Log

Animal Name / ID/ Group ______________________________
Location / Pen / Pasture ________________________________
Date ________________
Feeding Time (Circle) Morning / Midday / Evening / Other
Feed Type ____________________
Amount Given ________________
Supplements / Medications _____________________________

____
Fed By ______________
Additional Notes _______________________________________

____

____

Animal Name / ID/ Group ______________________________
Location / Pen / Pasture ________________________________
Date ________________
Feeding Time (Circle) Morning / Midday / Evening / Other
Feed Type ____________________
Amount Given ________________
Supplements / Medications _____________________________

____
Fed By ______________
Additional Notes _______________________________________

____

____

Feeding Log

Animal Name / ID/ Group ______________________________

Location / Pen / Pasture ________________________________

Date ________________

Feeding Time (Circle) Morning / Midday / Evening / Other

Feed Type ____________________

Amount Given ________________

Supplements / Medications _____________________________

____

Fed By ______________

Additional Notes _______________________________________

____

____

Animal Name / ID/ Group ______________________________

Location / Pen / Pasture ________________________________

Date ________________

Feeding Time (Circle) Morning / Midday / Evening / Other

Feed Type ____________________

Amount Given ________________

Supplements / Medications _____________________________

____

Fed By ______________

Additional Notes _______________________________________

____

____

Feeding Log

Animal Name / ID/ Group ______________________________

Location / Pen / Pasture ________________________________

Date _______________

Feeding Time (Circle) Morning / Midday / Evening / Other

Feed Type ____________________

Amount Given _______________

Supplements / Medications _____________________________

____

Fed By _____________

Additional Notes _____________________________________

____

____

Animal Name / ID/ Group ______________________________

Location / Pen / Pasture ________________________________

Date _______________

Feeding Time (Circle) Morning / Midday / Evening / Other

Feed Type ____________________

Amount Given _______________

Supplements / Medications _____________________________

____

Fed By _____________

Additional Notes _____________________________________

____

____

Feeding Log

Animal Name / ID/ Group ______________________________

Location / Pen / Pasture ________________________________

Date ________________

Feeding Time (Circle) Morning / Midday / Evening / Other

Feed Type ____________________

Amount Given ________________

Supplements / Medications _____________________________

____

Fed By ______________

Additional Notes _______________________________________

____

____

Animal Name / ID/ Group ______________________________

Location / Pen / Pasture ________________________________

Date ________________

Feeding Time (Circle) Morning / Midday / Evening / Other

Feed Type ____________________

Amount Given ________________

Supplements / Medications _____________________________

____

Fed By ______________

Additional Notes _______________________________________

____

____

Feeding Log

Animal Name / ID/ Group ______________________________
Location / Pen / Pasture ________________________________
Date ________________
Feeding Time (Circle) Morning / Midday / Evening / Other
Feed Type ____________________
Amount Given ________________
Supplements / Medications _____________________________

____
Fed By ______________
Additional Notes _______________________________________

____

____

Animal Name / ID/ Group ______________________________
Location / Pen / Pasture ________________________________
Date ________________
Feeding Time (Circle) Morning / Midday / Evening / Other
Feed Type ____________________
Amount Given ________________
Supplements / Medications _____________________________

____
Fed By ______________
Additional Notes _______________________________________

____

____

Feeding Log

Animal Name / ID/ Group ______________________________
Location / Pen / Pasture ________________________________
Date ________________
Feeding Time (Circle) Morning / Midday / Evening / Other
Feed Type ____________________
Amount Given ________________
Supplements / Medications ____________________________
____
Fed By ______________
Additional Notes ______________________________________
____
____

Animal Name / ID/ Group ______________________________
Location / Pen / Pasture ________________________________
Date ________________
Feeding Time (Circle) Morning / Midday / Evening / Other
Feed Type ____________________
Amount Given ________________
Supplements / Medications ____________________________
____
Fed By ______________
Additional Notes ______________________________________
____
____

Feeding Log

Animal Name / ID/ Group ______________________________
Location / Pen / Pasture ________________________________
Date ________________
Feeding Time (Circle) Morning / Midday / Evening / Other
Feed Type ____________________
Amount Given ________________
Supplements / Medications _____________________________

____

Fed By ______________
Additional Notes ______________________________________

____

____

Animal Name / ID/ Group ______________________________
Location / Pen / Pasture ________________________________
Date ________________
Feeding Time (Circle) Morning / Midday / Evening / Other
Feed Type ____________________
Amount Given ________________
Supplements / Medications _____________________________

____

Fed By ______________
Additional Notes ______________________________________

____

____

Feeding Log

Animal Name / ID/ Group ____________________________
Location / Pen / Pasture ______________________________
Date _______________
Feeding Time (Circle) Morning / Midday / Evening / Other
Feed Type ___________________
Amount Given _______________
Supplements / Medications ___________________________

____
Fed By _____________
Additional Notes ____________________________________

____

____

Animal Name / ID/ Group ____________________________
Location / Pen / Pasture ______________________________
Date _______________
Feeding Time (Circle) Morning / Midday / Evening / Other
Feed Type ___________________
Amount Given _______________
Supplements / Medications ___________________________

____
Fed By _____________
Additional Notes ____________________________________

____

____

Feeding Log

Animal Name / ID/ Group ______________________________
Location / Pen / Pasture ________________________________
Date ________________
Feeding Time (Circle) Morning / Midday / Evening / Other
Feed Type ____________________
Amount Given ________________
Supplements / Medications _____________________________

____

Fed By ______________
Additional Notes _______________________________________

____

____

Animal Name / ID/ Group ______________________________
Location / Pen / Pasture ________________________________
Date ________________
Feeding Time (Circle) Morning / Midday / Evening / Other
Feed Type ____________________
Amount Given ________________
Supplements / Medications _____________________________

____

Fed By ______________
Additional Notes _______________________________________

____

____

Feeding Log

Animal Name / ID/ Group ______________________________
Location / Pen / Pasture ________________________________
Date ________________
Feeding Time (Circle) Morning / Midday / Evening / Other
Feed Type ____________________
Amount Given ________________
Supplements / Medications ____________________________

____
Fed By ______________
Additional Notes ______________________________________
____
____

Animal Name / ID/ Group ______________________________
Location / Pen / Pasture ________________________________
Date ________________
Feeding Time (Circle) Morning / Midday / Evening / Other
Feed Type ____________________
Amount Given ________________
Supplements / Medications ____________________________

____
Fed By ______________
Additional Notes ______________________________________
____
____

Feeding Log

Animal Name / ID/ Group ______________________________
Location / Pen / Pasture ________________________________
Date ________________
Feeding Time (Circle) Morning / Midday / Evening / Other
Feed Type ____________________
Amount Given ________________
Supplements / Medications _____________________________

____
Fed By ______________
Additional Notes ______________________________________
____
____

Animal Name / ID/ Group ______________________________
Location / Pen / Pasture ________________________________
Date ________________
Feeding Time (Circle) Morning / Midday / Evening / Other
Feed Type ____________________
Amount Given ________________
Supplements / Medications _____________________________

____
Fed By ______________
Additional Notes ______________________________________
____
____

Feeding Log

Animal Name / ID/ Group ______________________________
Location / Pen / Pasture ________________________________
Date ________________
Feeding Time (Circle) Morning / Midday / Evening / Other
Feed Type ____________________
Amount Given ________________
Supplements / Medications _____________________________

____
Fed By ______________
Additional Notes ____
____
____

Animal Name / ID/ Group ______________________________
Location / Pen / Pasture ________________________________
Date ________________
Feeding Time (Circle) Morning / Midday / Evening / Other
Feed Type ____________________
Amount Given ________________
Supplements / Medications _____________________________

____
Fed By ______________
Additional Notes ____
____
____

Feeding Log

Animal Name / ID/ Group ______________________________
Location / Pen / Pasture ________________________________
Date _______________
Feeding Time (Circle) Morning / Midday / Evening / Other
Feed Type ____________________
Amount Given ________________
Supplements / Medications _____________________________

____
Fed By ______________
Additional Notes _______________________________________
____
____

Animal Name / ID/ Group ______________________________
Location / Pen / Pasture ________________________________
Date _______________
Feeding Time (Circle) Morning / Midday / Evening / Other
Feed Type ____________________
Amount Given ________________
Supplements / Medications _____________________________

____
Fed By ______________
Additional Notes _______________________________________
____
____

Feeding Log

Animal Name / ID/ Group ______________________________
Location / Pen / Pasture _________________________________
Date ________________
Feeding Time (Circle) Morning / Midday / Evening / Other
Feed Type _____________________
Amount Given ________________
Supplements / Medications _____________________________
____
Fed By ______________
Additional Notes ______________________________________
____
____

Animal Name / ID/ Group ______________________________
Location / Pen / Pasture _________________________________
Date ________________
Feeding Time (Circle) Morning / Midday / Evening / Other
Feed Type _____________________
Amount Given ________________
Supplements / Medications _____________________________
____
Fed By ______________
Additional Notes ______________________________________
____
____

Feeding Log

Animal Name / ID/ Group ______________________________
Location / Pen / Pasture ________________________________
Date ________________
Feeding Time (Circle) Morning / Midday / Evening / Other
Feed Type ____________________
Amount Given ________________
Supplements / Medications _____________________________

____
Fed By ______________
Additional Notes ______________________________________
____
____

Animal Name / ID/ Group ______________________________
Location / Pen / Pasture ________________________________
Date ________________
Feeding Time (Circle) Morning / Midday / Evening / Other
Feed Type ____________________
Amount Given ________________
Supplements / Medications _____________________________

____
Fed By ______________
Additional Notes ______________________________________
____
____

Feeding Log

Animal Name / ID/ Group ______________________________
Location / Pen / Pasture ________________________________
Date ________________
Feeding Time (Circle) Morning / Midday / Evening / Other
Feed Type ____________________
Amount Given ________________
Supplements / Medications _____________________________

____
Fed By ______________
Additional Notes _______________________________________

____

____

Animal Name / ID/ Group ______________________________
Location / Pen / Pasture ________________________________
Date ________________
Feeding Time (Circle) Morning / Midday / Evening / Other
Feed Type ____________________
Amount Given ________________
Supplements / Medications _____________________________

____
Fed By ______________
Additional Notes _______________________________________

____

____

Feeding Log

Animal Name / ID/ Group ______________________________
Location / Pen / Pasture ________________________________
Date ________________
Feeding Time (Circle) Morning / Midday / Evening / Other
Feed Type ____________________
Amount Given ________________
Supplements / Medications ______________________________

____
Fed By ______________
Additional Notes ____
____
____

Animal Name / ID/ Group ______________________________
Location / Pen / Pasture ________________________________
Date ________________
Feeding Time (Circle) Morning / Midday / Evening / Other
Feed Type ____________________
Amount Given ________________
Supplements / Medications ______________________________

____
Fed By ______________
Additional Notes ____
____
____

Feeding Log

Animal Name / ID/ Group ______________________
Location / Pen / Pasture ______________________
Date ____________
Feeding Time (Circle) Morning / Midday / Evening / Other
Feed Type ________________
Amount Given ____________
Supplements / Medications ______________________

____

Fed By __________
Additional Notes ______________________________

____

____

Animal Name / ID/ Group ______________________
Location / Pen / Pasture ______________________
Date ____________
Feeding Time (Circle) Morning / Midday / Evening / Other
Feed Type ________________
Amount Given ____________
Supplements / Medications ______________________

____

Fed By __________
Additional Notes ______________________________

____

____

Hay & Feed Inventory Record

Animal / Group / Location Area ______________________________

Feed / Hay Type ______________________________

Feed / Hay Type ______________________________

Quantity Purchased ______________________________

Date Purchased ______________________________

Supplier # 1 ______________________________

Supplier # 2 ______________________________

Supplier # 3 ______________________________

Supplier	Weekly Total	Monthly Total	Yearly Total
	Total:	Total:	Total:

Animal / Group / Location Area ______________________________

Feed / Hay Type ______________________________

Feed / Hay Type ______________________________

Quantity Purchased ______________________________

Date Purchased ______________________________

Supplier # 1 ______________________________

Supplier # 2 ______________________________

Supplier # 3 ______________________________

Supplier	Weekly Total	Monthly Total	Yearly Total
	Total:	Total:	Total:

Hay & Feed Inventory Record

Animal / Group / Location Area ______________________

Feed / Hay Type ______________________

Feed / Hay Type ______________________

Quantity Purchased ______________________

Date Purchased ______________________

Supplier # 1 ______________________

Supplier # 2 ______________________

Supplier # 3 ______________________

Supplier	Weekly Total	Monthly Total	Yearly Total
	Total:	Total:	Total:

Animal / Group / Location Area ______________________

Feed / Hay Type ______________________

Feed / Hay Type ______________________

Quantity Purchased ______________________

Date Purchased ______________________

Supplier # 1 ______________________

Supplier # 2 ______________________

Supplier # 3 ______________________

Supplier	Weekly Total	Monthly Total	Yearly Total
	Total:	Total:	Total:

Hay & Feed Inventory Record

Animal / Group / Location Area ______________________

Feed / Hay Type ______________________

Feed / Hay Type ______________________

Quantity Purchased ______________________

Date Purchased ______________________

Supplier # 1 ______________________

Supplier # 2 ______________________

Supplier # 3 ______________________

Supplier	Weekly Total	Monthly Total	Yearly Total
	Total:	Total:	Total:

Animal / Group / Location Area ______________________

Feed / Hay Type ______________________

Feed / Hay Type ______________________

Quantity Purchased ______________________

Date Purchased ______________________

Supplier # 1 ______________________

Supplier # 2 ______________________

Supplier # 3 ______________________

Supplier	Weekly Total	Monthly Total	Yearly Total
	Total:	Total:	Total:

Hay & Feed Inventory Record

Animal / Group / Location Area ______________________
Feed / Hay Type ______________________
Feed / Hay Type ______________________
Quantity Purchased ______________________
Date Purchased ______________________
Supplier # 1 ______________________
Supplier # 2 ______________________
Supplier # 3 ______________________

Supplier	Weekly Total	Monthly Total	Yearly Total
	Total:	Total:	Total:

Animal / Group / Location Area ______________________
Feed / Hay Type ______________________
Feed / Hay Type ______________________
Quantity Purchased ______________________
Date Purchased ______________________
Supplier # 1 ______________________
Supplier # 2 ______________________
Supplier # 3 ______________________

Supplier	Weekly Total	Monthly Total	Yearly Total
	Total:	Total:	Total:

Hay & Feed Inventory Record

Animal / Group / Location Area ______________________________

Feed / Hay Type ______________________________

Feed / Hay Type ______________________________

Quantity Purchased ______________________________

Date Purchased ______________________________

Supplier # 1 ______________________________

Supplier # 2 ______________________________

Supplier # 3 ______________________________

Supplier	Weekly Total	Monthly Total	Yearly Total
	Total:	Total:	Total:

Animal / Group / Location Area ______________________________

Feed / Hay Type ______________________________

Feed / Hay Type ______________________________

Quantity Purchased ______________________________

Date Purchased ______________________________

Supplier # 1 ______________________________

Supplier # 2 ______________________________

Supplier # 3 ______________________________

Supplier	Weekly Total	Monthly Total	Yearly Total
	Total:	Total:	Total:

Hay & Feed Inventory Record

Animal / Group / Location Area ______________________

Feed / Hay Type ______________________

Feed / Hay Type ______________________

Quantity Purchased ______________________

Date Purchased ______________________

Supplier # 1 ______________________

Supplier # 2 ______________________

Supplier # 3 ______________________

Supplier	Weekly Total	Monthly Total	Yearly Total
	Total:	Total:	Total:

Animal / Group / Location Area ______________________

Feed / Hay Type ______________________

Feed / Hay Type ______________________

Quantity Purchased ______________________

Date Purchased ______________________

Supplier # 1 ______________________

Supplier # 2 ______________________

Supplier # 3 ______________________

Supplier	Weekly Total	Monthly Total	Yearly Total
	Total:	Total:	Total:

Hay & Feed Inventory Record

Animal / Group / Location Area ______________________________

Feed / Hay Type ______________________________

Feed / Hay Type ______________________________

Quantity Purchased ______________________________

Date Purchased ______________________________

Supplier # 1 ______________________________

Supplier # 2 ______________________________

Supplier # 3 ______________________________

Supplier	Weekly Total	Monthly Total	Yearly Total
	Total:	Total:	Total:

Animal / Group / Location Area ______________________________

Feed / Hay Type ______________________________

Feed / Hay Type ______________________________

Quantity Purchased ______________________________

Date Purchased ______________________________

Supplier # 1 ______________________________

Supplier # 2 ______________________________

Supplier # 3 ______________________________

Supplier	Weekly Total	Monthly Total	Yearly Total
	Total:	Total:	Total:

Hay & Feed Inventory Record

Animal / Group / Location Area ____________________

Feed / Hay Type ____________________

Feed / Hay Type ____________________

Quantity Purchased ____________________

Date Purchased ____________________

Supplier # 1 ____________________

Supplier # 2 ____________________

Supplier # 3 ____________________

Supplier	Weekly Total	Monthly Total	Yearly Total
	Total:	Total:	Total:

Animal / Group / Location Area ____________________

Feed / Hay Type ____________________

Feed / Hay Type ____________________

Quantity Purchased ____________________

Date Purchased ____________________

Supplier # 1 ____________________

Supplier # 2 ____________________

Supplier # 3 ____________________

Supplier	Weekly Total	Monthly Total	Yearly Total
	Total:	Total:	Total:

Farrier Yearly Cost Record

Farrier Name ______________________________

Year __________

Month	Date Paid	Amount Paid
January		
February		
March		
April		
May		
June		
July		
August		
September		
October		
November		
December		

Yearly Total ____________

Farrier Yearly Cost Record

Farrier Name ______________________________

Year __________

Month	Date Paid	Amount Paid
January		
February		
March		
April		
May		
June		
July		
August		
September		
October		
November		
December		

Yearly Total ____________

Farrier Yearly cost Record

Farrier Name ______________________________

Year __________

Month	Date Paid	Amount Paid
January		
February		
March		
April		
May		
June		
July		
August		
September		
October		
November		
December		

Yearly Total ____________

Veterinarian Expense Record

Vet Clinic / Veterinarian ______________________________

____

Year ____________

Date	Animal /Group	Service	Cost

Yearly Total Vet Cost ____________

Veterinarian Expense Record

Vet Clinic / Veterinarian ______________________________

____

Year ____________

Date	Animal /Group	Service	Cost

Yearly Total Vet Cost ____________

Veterinarian Expense Record

Vet Clinic / Veterinarian ______________________________

____

Year ____________

Date	Animal /Group	Service	Cost

Yearly Total Vet Cost ____________

Medication & Supplement Expense Record

Date	Animal / Group	Medication / Supplement Cost

Yearly Total Medication & Supplement Cost ________________

Medication & Supplement Expense Record

Date	Animal / Group	Medication / Supplement Cost

Yearly Total Medication & Supplement Cost _______________

Medication & Supplement Expense Record

Date	Animal / Group	Medication / Supplement Cost

Yearly Total Medication & Supplement Cost ______________

General Farm Expense Record

Date	Description	Cost

Yearly Total General Farm Expense ___________

General Farm Expense Record

Date	Description	Cost

Yearly Total General Farm Expense ____________

General Farm Expense Record

Date	Description	Cost

Yearly Total General Farm Expense ____________

Annual Farm Expense Summary

Business /Organization Name ____________________________
Year _____________

Expense Category	Annual Total
Feed / Hay	
Farrier	
Veterinary	
Medication & Supplements	
General Farm Expenses	

Total Annual Farm Expense ______________

Annual Farm Expense Summary

Business /Organization Name ______________________________
Year ______________

Expense Category	Annual Total
Feed / Hay	
Farrier	
Veterinary	
Medication & Supplements	
General Farm Expenses	

Total Annual Farm Expense _______________

Annual Farm Expense Summary

Business /Organization Name ______________________________
Year ______________

Expense Category	Annual Total
Feed / Hay	
Farrier	
Veterinary	
Medication & Supplements	
General Farm Expenses	

Total Annual Farm Expense $ _______________

Notes

Animal Loss / Deceased Record

Date Deceased	Animal Name / ID	Species	Cause of Death

Animal Loss / Deceased Record

Date Deceased	Animal Name / ID	Species	Cause of Death

Animal Loss / Deceased Record

Date Deceased	Animal Name / ID	Species	Cause of Death

ISBN: 978-1-971331-06-5

This book is intended as a livestock and farm, ranch, sanctuary, rescue, and homestead use. The author and publisher assume no responsibility for errors or omissions, or for the results obtained from the use of the information contained in this book.

Interior design by Angie Cservak.

Printed in the United States of America.

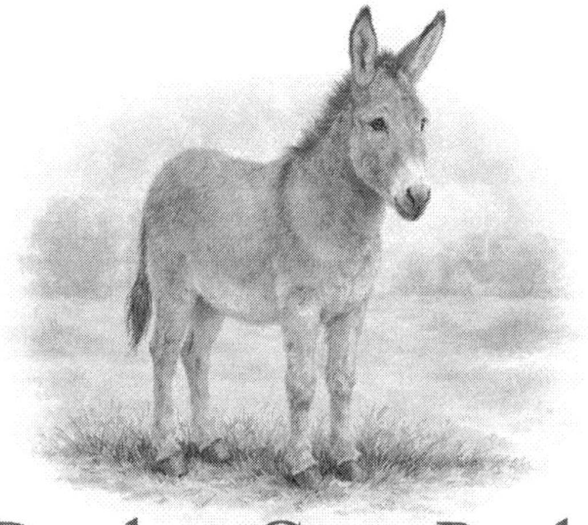

Donkey Gene Books

Discover More from Donkey Gene Books

Scan the QR code to explore more books and stories from the Donkey Gene collection.

Visit to see available children's farm stories featuring donkeys, horses, and life on the farm.

Made in the USA
Monee, IL
08 November 2026

68076596R00196